The Family Dynamic

The Family Dynamic

A Journey into the Mystery
of Sibling Success

Susan Dominus

CROWN
NEW YORK

CROWN

An imprint of the Crown Publishing Group
A division of Penguin Random House LLC
1745 Broadway
New York, NY 10019
crownpublishing.com
penguinrandomhouse.com

Library of Congress Cataloging-in-Publication Data
Names: Dominus, Susan, author. Title: The Family Dynamic: A Journey into the Mystery of Sibling Success / by Susan Dominus. Description: New York: Crown Publishing, [2025] | Includes bibliographical references. Identifiers: LCCN 2023058925 (print) | LCCN 2023058926 (ebook) | ISBN 9780593137901 (hardcover) | ISBN 9780593137918 (ebook) | Subjects: LCSH: Families. | Children. | Parenting. | Behavior genetics. Classification: LCC HQ755.8 .D653 2024 (print) | LCC HQ755.8 (ebook) | DDC 306.85—dc23/eng/20240402
LC record available at https://lccn.loc.gov/2023058925
LC ebook record available at https://lccn.loc.gov/2023058926

Hardback ISBN 978-0-593-13790-1
Ebook ISBN 978-0-593-13791-8

Editor: Gillian Blake
Editorial assistant: Amy Li
Production editor: Ashley Pierce
Text designer: Aubrey Khan
Production: Heather Williamson
Copy editors: Hilary Roberts and Diana Drew
Proofreaders: Tracy Lynch and Alissa Fitzgerald
Publicist: Gwyneth Stansfield
Marketer: Kimberly Lew

Manufactured in the United States of America

9 8 7 6 5 4 3 2 1

First Edition

The authorized representative in the EU for product safety and compliance is Penguin Random House Ireland, Morrison Chambers, 32 Nassau Street, Dublin D02 YH68, Ireland, https://eu-contact.penguin.ie.

To my boys

Contents

Introduction

I have been reporting this book, in a sense, since I was in fourth grade. That's how old I was when my parents went on one of many two-week trips required for my father's work, leaving me with close family friends who lived in a rambling three-story Victorian. The house was located just a few blocks from our own, but its rites and rhythms were distinct from ours. In our house, members of our family could reliably be found after dinner enjoying '80s-era sitcoms in a small den; I don't remember what our family friends' three sons did during those hours before bed, but I know the television, located in a room off the kitchen, went untouched. At night, their mom, Aunt Gail, as I called her, often sat in that room, reading, all but stone-still, well past midnight, with a lamp shining brightly on whatever heavy book lay on her lap.

The difference in family habits was even more apparent at dinnertime, when my mother and father generally talked in shorthand about my father's work life, while my brother, my sister, and I typically wolfed down our meals, goofed around, or debated once more the question of why my sister always got the corner seat. In the household where I was a guest, after the family enjoyed what seemed to me to be especially elegant meals, the father would put down his fork, look up at one of his three sons, and either ask him about a particular point of current events or pose a math problem invented on the spot. A plane was leaving from one city at 7 A.M., going so many miles an hour, just as another plane was departing from another city at another given time, traveling at some other high speed. At what point would they meet? The sons were expected to work this answer out in their head, a ritual of mathematical sport I enjoyed to no end until the evening when the father posed such a question to me. After an interminable silence, my mind entirely blank, I burst into tears.

As the only daughter figure in the house, I was indulged and even a little spoiled at our family friends' house, but I still felt considerable relief when my parents returned. I reverted to the more manageable mealtime challenges my parents posed: cleaning my plate and chewing with my mouth closed. At the same time, I did find myself reflecting on other possibilities of which I was now aware. I had been given a window into just how varied family culture could be regarding expectations and the cultivation of skills. Math did not come easily to me—but what if I'd been working problems out around the dinner table night after night? What if I'd become accustomed to voicing my opinion on the events of the day, defending my stance—where would that have taken me? Were those kinds of expectations a blessing and a gift or a burden that came with a steady low-grade pressure?

As I grew up, I became a bit of a family-ologist, paying close attention to the clues and details of my friends' upbringings, interested—especially—in how they approached enrichment at home. Even before I met my dear friend Anya Epstein our sophomore year of college, I knew, from mutual friends, her family pedigree—that her father, Leslie Epstein, was an esteemed novelist; that her grandfather, Philip Epstein, had co-written, with his twin brother, Julius, *Casablanca,* among other cinematic gems. I absorbed with fascination, after Anya and I became roommates in college, the stories of a household where expectations were high for the children but also for the arts. At one play she attended near her home in Brookline, Massachusetts, Anya, then nine, leaped up to join the rest of the audience in a standing ovation. "Boston audiences," Leslie hissed, tugging at her skirt to pull her back into her seat. Her father read Dickens to her at bedtime; John Updike and Doris Lessing joined the family when they entertained. In some of her earliest, fondest memories with her father, the two are talking about ideas for a children's story about starfish and the stars that Epstein was thinking about trying to publish.

I was curious about other elements of their family culture that seemed bound up with their many accomplishments—a certain confidence or even boldness. The family celebrated irreverence, practical jokes, a discreet but well-placed thumb to the nose in the name of justice. The family took considerable pride in how the twins responded when the House Un-American Activities Committee subpoenaed them after their boss, Jack Warner, one of the founders of Warner Bros., included their names in a list of writers he suspected of being communist sympathizers. An official questionnaire they received asked, as Leslie Epstein relayed in *Tablet* magazine, if they'd ever been a member of a subversive organization, and if so, which one. "To the former, the Epsteins answered: Yes. To the latter, the reply was: Warner Bros.," Leslie Epstein wrote. "They never heard from the committee again."

The family reveled in the pleasures of wit and language and, except for Anya, sports; some of Leslie's warmest recollections of his uncle and his father, who died tragically young, were of football games they attended together. Anya's father was unreachable to her only when he was occupied with one of two endeavors: when he was writing, and when he was watching the Red Sox with her younger twin brothers.

Narrative was even at the heart of the success of Anya's mother, Ilene Epstein, a warm, psychologically intuitive woman, who owned, along with her own twin and a close friend, a clothing store called The Studio, in Brookline, a local institution for some forty years. Its loyal customers came to shop, but also to sit and share story after story, many of which were relayed to Anya over dinner.

Over time, I have watched the Epstein children's lives unfold in ways that seemed so clearly to reflect their upbringing—their many advantages, to be sure, but also the family's values, influences, and expectations. Anya has a thriving career writing for television and served as a showrunner on shows like *The Affair* and *In Treatment*. One of her younger brothers, Paul, is a nonprofit founder and beloved social worker with the Brookline public school system. In 2002, his twin, Theo Epstein, became the youngest general manager of a baseball team, the Red Sox, at age twenty-eight, a position from which he launched one of the most illustrious careers in modern baseball history. He led both the Red Sox and then the Chicago Cubs to historic triumphs; he landed on the cover of *Fortune* and was the number one slot on its 2017 "World's Greatest Leaders" list. Theo, now a partial owner of the Red Sox and a senior adviser to the Fenway Sports Group, and Paul, through a foundation they started together, have raised more than $16 million to fund college tuition for disadvantaged students and support nonprofits in Boston and Chicago.

"It's a little dizzying—that's a good word," Leslie Epstein conceded to *The Boston Globe* when Theo, at twenty-eight, took the job as general manager of the Red Sox. "But at Theo's age, Alexander the Great was already general manager of the world."

Even before I met Anya, I was especially attuned to stories that I came across about the childhood rituals of enrichment that popped up in the lore of various famously successful families, such as the Kennedy family, which so clearly prized both education and competition. In her memoir, *My Twelve Years with John F. Kennedy*, Evelyn Lincoln, JFK's personal secretary, recalled that Kennedy's "inquiring mind and search for all the facts reflected, it seemed, the training he received around the family table when he was growing up." His father, Joseph P. Kennedy, Sr., would assign a topic—in her book, Lincoln gave the example of "Algeria"—that one child would be expected to present with some expertise to the group over dinner. But all the other children in the family were expected to prepare on that subject, too, "so they could question the first one when he made his report and see how much he really knew," wrote Lincoln.

I was so curious about these kinds of dinnertime rituals in certain kinds of high-profile families that in 2012, I wrote an article for which I interviewed people such as Franklin Foer, one of three brothers who have all written best-selling and acclaimed books. When I spoke to him, I recalled an article in *The New York Observer* in which Franklin, now a staff writer at *The Atlantic*, was asked if it was true that the brothers had also been asked to present on various topics at the dinner table. "That was only me, 'cause I was shy," Franklin Foer clarified to the *Observer*. But he told me, when I spoke to him for my article, that his family's dinner-table conversation was perhaps unusually wide-ranging and open-minded. His father (I wrote) "led his children in debates about economic policy and civil rights issues, but with an open ear: a conversation about Reagan's Star Wars policies might lead to a discussion about 'why we couldn't build a giant shield over the United States out of Legos.'" Scatological humor was welcome; the family culture was expansive, with a "sky's-the-limit" sensibility. And, yes, great optimism—some might say ambition—was clearly encouraged, Franklin told me in a subsequent interview: The Foers' mother, who wanted to foster self-confidence in her children, created albums for them on their birthdays in which their heads were transposed onto the bodies of great figures whom they admired—Han Solo, the football player Sid Luckman, and Thomas Jefferson.

ALTHOUGH THIS ONGOING INTEREST of mine was sparked by that early-childhood exposure to the home of family friends, I can also trace my fascination with notable families to one I never met, although I felt, through hours of reading and rereading, that I came to know them well: the Brontës, a nineteenth-century trio of sisters whose novels are still devoured, dissected, and revered by lovers of literature. Even as I wept over the indignities suffered by the title character in Charlotte Brontë's *Jane Eyre* or was thrilled by the cruel nature of passion in her sister Emily's *Wuthering Heights,* some separate part of me already was noodling over what would become a familiar preoccupation: What were the circumstances that created two sisters who wrote books such as those? It took me a bit longer to come to the work of the family's youngest sibling, Anne, at which point my curiosity only grew. Anne Brontë's novels—*Agnes Grey* and *The Tenant of Wildfell Hall*—have never become cultural touchstones as *Jane Eyre* and *Wuthering Heights* have, but Anne, too, wrote books that stand on their own as originals, feminist cris de coeur that are unflinching and innovative. What exactly had transpired in the Brontë household, what family climate had evolved, so that lightning struck three times?

In 1857, two years after Charlotte died at age thirty-eight, her friend and fellow novelist, Elizabeth Gaskell, published *The Life of Charlotte Brontë,* a biography. In its own way, Gaskell's book tried to answer the same questions at the heart of each of the chapters of this book: How did a family as anomalous as this one emerge? Gaskell's answers (which have also been perceived as excuses for female novelists who wrote such "coarse" works) were intriguing enough that the book was immediately a critical and financial success, the mother of countless biographies, films, and television series, each developed to satisfy the curiosities of the next generation. A biography of Emily, published in 1883, presented a reverent critical reappraisal of her work, but also delved into her upbringing, suggesting just how much the family lore fed into appreciation for the sisters' work.

At some point, I realized that I was interested enough in exceptional families that I wanted to read a book that explored the themes common to many of them—and since I was not aware of one, I would have to research it and write it myself. As I collected details and noted recurring phenomena, I kept reflecting on various facets of the Brontë family story, which was rich and complicated and sprawling enough to sustain all those fascinated Brontë-ologists—but also to include so many of the salient themes that eventually emerged in my own reporting on several contemporary families. The Brontë family fascinates, not just because of its brilliance, I would argue, but because of the elements in their story that resonate, so much so that aspects of their upbringing became touchstones to me, throwing into high relief details of other families' stories I might not have otherwise recognized as notable.

In many of the families whose stories I explored, I saw, for example, different iterations of Patrick Brontë, a widowed minister, a father endowed—for better or for worse—with so much strength that he outlived not just his wife, but eventually, all six of his children. I came to think of Patrick Brontë as an overcomer, an against-the-odds-er, someone with the kind of story that burrows into a family's sense of self and history. Those parents are, within the family, living legends, even if they never reach any kind of unusual renown outside of it; they provide living proof that other people's limits need not apply to their family.

Those kinds of parents may be vocally ambitious for their children, as they were for themselves or, without ever saying anything explicit, they may create a culture of reaching simply by having set the bar high in their own path. They unleash an arrow of ambition, and are often surprised to find that they can't control that arrow's direction. Patrick Brontë loved poetry and fiction, just as his daughters did, but he was too much a man of his time to desire that any of them pursue dreams of literary greatness. If anything, Charlotte implied in a letter, her father encouraged her to focus on "the duties a woman ought to fulfill"— tasks like sewing or teaching, not reading or writing. Once Charlotte was a star novelist, however, he sent a copy of *Jane Eyre* to his relatives in Ireland, inscribed with a note brimming with pride (although the book was published under a pseudonym, her real name, he made sure they knew, "is everywhere known").

The Brontës helped me see that the siblings are often the ones who help guide the direction of the arrow into the future. Charlotte was the driving force of her sisters' professional success, an accidental talent scout who saw possibility and seized on it. In the fall of 1845, at age twenty-nine, Charlotte, she later wrote in an essay, stumbled on pages of poetry in her sister Emily's handwriting—work that carried a "peculiar music—wild, melancholy, and elevating." At first, Emily was annoyed by Charlotte's snooping, with what she considered the "unwarrantable liberty" her older sister had taken in poking around her papers. "It took hours to reconcile her to the discovery I had made, and days to persuade her that such poems merited publication," Charlotte wrote in that essay, which would serve as the preface for an 1850 edition of *Wuthering Heights*. At some point, Anne offered up her own poems to Charlotte. The verdict? They had, "a sweet sincere pathos of their own," her older sister Charlotte wrote, not quite damning with faint praise. Charlotte convinced her sisters that they should, together, submit to editors a volume of their poetry, a venture that met with poor sales but that jump-started their lives as published writers.

Alone, any one of the Brontë sisters might have lost faith in her writing or kept her work private—but they had, in one another, when they were writing those novels, an artists' colony of three, populated exclusively by women who supported one another's work.

Through the Brontës, I came to see that my own book would not just be about sibling influences or parental inspirations—it would have to be about the family dynamics in these dynamic families, the complicated interplay of various factions and influences. One can analyze the elements that led to the Brontës' remarkable achievements with fascination, without expecting ever to see them replicated in full. As a family, they represent, together, one of the world's most beautiful statistical anomalies, and I will return to them several times in these chapters.

WHEN I STARTED THIS BOOK, I was the mother of young twin boys, now young men at the start of their college careers. Throughout that span of time, my husband and I would (fitfully) try to be thoughtful about our own family culture, our own expectations and influences. I was not looking to foster towering ambition, but I worried about how the choices we were making might shape our sons' futures, their ability to enjoy the full range of their potential. As a young person, I had been enormously grateful for the healthy distance my parents maintained from my own intensely high academic aspirations. But now that I was a parent myself, as my children passed from one grade to the next, I found myself slipping into the parenting anxieties common to a certain demographic in our time: Were we exposing them to enough, encouraging them enough, enriching them enough? Were we instilling in them a strong enough work ethic?

Most parents would say that what they care about chiefly is their children's general and future happiness; I am among them. And yet somehow, most parents do not see it as their charge to make a reasonably happy child even happier. They do not buy books with titles such as, say, *How to Turn Your Happy Child into an Ecstatic One*. Most parents do not try to quantify whether their child is in the top 10 percent of happy children in the grade, nor do the parents of that reasonably content young person worry, *Perhaps I did something wrong—I did not provide enough opportunities for happiness*. They do not think, *Unless I make my child even happier, my child will not thrive*.

Achievement, on the other hand, can be a competitive parental sport; it comes with handy and concrete metrics—grades, scores, gold stars, scholarships, rankings. Parents with a certain mindset may feel they can never rest. If a daughter is acing her math class, those parents might wonder: Is the math class not challenging enough? If the son is not picking up his oboe to practice regularly, the parents might wonder: Should I try rewards? (Behavioral scientists say yes.) But if I do reward certain behaviors, won't that undermine my child's intrinsic motivation? (Developmental psychologists say yes.)

Parents also must contend with a feeling of responsibility, the sense that if a child does not possess some of the qualities our society values—conscientiousness, for instance—maybe they have failed to emphasize the right things. They have not encouraged them at the right moment or followed through on warned consequences. If only they had parented better, surely their children would be succeeding more demonstrably in the ways that schools and workplaces, especially, like them to.

Perhaps like me, you have looked at some families that have more than one high-achieving child, that seem to have some sort of streak going on at home, and wondered: *How did they do it? What was going on at the dinner table in that house? Is there anything to be gleaned from how they inspired their children?* These questions are not necessarily about ambition for ambition's sake. For many parents contemplating these questions, the goal is to ensure that their children feel agency or feel equipped to reach higher—to link achievement not to personal glory but to a larger purpose and meaning. They do not want their children to feel about their upbringing the way a friend once told me, with great sadness, that he did: that his parents never expected much of him, that he was alone with his dreams, which made reaching them a harder lift.

About my own twin boys, when I think, *I just want them to be happy,* perhaps what I really mean is, *I want them to be happy now and eventually.* That starts with unconditional love, knowing that they have it and being capable of giving it. But another source of happiness lies in a sense of one's agency—and so I also hope that they will have been raised to be curious and resourceful, capable of contributing, in the future, to whatever callings they pursue.

Probably because my twins are fraternal, I am especially interested in consistency of outcomes in families. I suspect that parents of twins, especially fraternal twins, spend more time than most thinking about fairness and equality—about their hope that not just one but both of their children will lead lives that are rich with relationships and that are also interesting and rewarding. The longing for some reliable (no doubt elusive) equalizing guidance is powerful.

When predicting how likely certain kinds of success are, sociological researchers typically factor in socioeconomic status, or SES. SES is the single most powerful predictor of how much education a family's child will attain, but sociologists note that great educational and economic disparities still exist between and among siblings from the same family, especially where resources are scarce. "In explaining economic inequality in America, sibling differences represent about three-quarters of all the differences between individuals," writes the sociologist Dalton Conley in his 2004 book, *The Pecking Order.*

Part of that discrepancy can be explained by an uneven distribution of luck, and part of that can be explained by the luck, specifically, of the genes: Although, on average, siblings are more alike than first cousins, and first cousins are more alike than second cousins, the genes can be shuffled in surprising ways that might yield dramatic differences among children in one nuclear family. How those genes shake out, collectively, can mean the difference between a child with ADHD, for whom school is a trial, and a child who likes the order of a classroom; genes might nudge one child toward books and another toward steady socializing. And those small differences, even early on, and depending on the environment, can set in motion a cascade of effects that result in different outcomes down the road.

Why siblings can have such different life courses is also rooted in the economics of their own family: Conley's research found that in middle-class and low-income families, parents tend to invest more heavily in the child who seems most destined for success, with tutoring or private school or college, leaving their less talented children with fewer resources with which to start out in life. (Wealthier families, by contrast, tend to spend their resources trying to help a struggling student succeed.) Families' financial ups and downs also fluctuate, so that siblings' outcomes can be a function of timing akin to musical chairs: Where were you in your educational process when the music stopped? Did the oldest sibling get to college before her parents' split threw off-kilter the possibility of future savings? Did the family business finally take off in time to get the youngest child to private school but too late for the older ones even to be interested in going back to college?

The discrepancies so common among siblings make it all the more remarkable when a family, especially a working-class or middle-class family, yields more than one child who achieves something of note. And so, I set out to find some—to see what, if any, family dynamics seemed to contribute to that series of successes.

It turned out to be a challenge to find families in which all the siblings agreed to participate; in some cases, other siblings vetoed even one sibling's intent to participate. Even in families whose siblings had high achievement in common, the brothers and sisters often had wildly different feelings about everything else, including privacy or their perspectives on their own families. I witnessed squabbling and mild insult-slinging among venerable public figures, as they worked out among themselves whether to participate (once that began, the answer was inevitably no). I saw one sibling's desire to talk thwarted by another, more senior sibling calling the shots. Occasionally, a sibling felt nervous about revealing a family secret, and I found myself an object of suspicion, even if I had no interest in that particular aspect of their shared lives. The intimacy of family life, the deeply ingrained loyalties and resentments among siblings that could never be spoken aloud, were all so hot to the touch. But, eventually, with gentle persistence, I was fortunate enough to find six generous families whose hold-out siblings, even if they did not want to cooperate, did not object. These families were willing to open up to me, to share family stories, to recount their own individual paths in ways that I tried to chart into some kind of map, sometimes noticing interesting intersections of those routes that they themselves had not perceived.

The families that fascinated me had in common what I think of as a culture of boldness—a belief that they could change the world for the better or create great art or break world records. To come to the

United States speaking no English at all but believing that you can nonetheless find a way to propel your children to great success, despite your own financial struggles—that, too, is an act of boldness.

When seeking out contemporary families for this book, I was drawn to those pursuing goals that require fierce commitment and a deep, abiding sense of certainty in oneself and one's pursuit: civil rights activists, artists and writers and musicians, entrepreneurs, judges, an Olympic athlete, a lauded and trailblazing director. I started out thinking I was writing a book about high-achieving families; in the end, I wrote a book about families that dream big but that also have the skills to act effectively to realize those dreams. The book also features stories of several of the scientists whose research is relevant to some of the questions I ask about nature versus nurture, expectations and motivation. More often than not, those scientists' own family backgrounds offered insights into why they were asking those questions, too.

It has been one of the greatest pleasures of my professional life to get to know the families whose stories I tell in this book. None of them pretends to share any secret of success; most of them are too modest, and it is hard, in any case, to see your family background with any perspective, as it was the air you breathed, the norm of all norms. I would say that although I did detect some common threads among these families—many of them shared by the Brontës—what shines through are the fascinating stories of unusual families who shared their complications, their struggles, their parents' failings and triumphs, and the sibling dynamics that pained them and propelled them. It is safe to say that if this is a book about high-achieving families, it is often also about the real costs of having that kind of drive, the sacrifices that having that kind of focus sometimes entails or even the emotional anguish. The people in this book did not endeavor to hide from me their personal or their family's flaws; to some degree, their stories are at least as likely to console those outside the stratosphere as they are to inspire them.

I wrote and reported this book over many years, which also means it is a book about the nature of fame and success. Had I started this book a year or two earlier, I might have tried to understand the root of the Weinstein brothers' becoming such outsized creative influences in the world of independent film. People whom I met at one point in their careers ended up, for better or for worse, in very different places by the time I finished my reporting many years later or experienced public scrutiny along the way. Instead of lionizing high-achieving families, this book more likely humanizes them.

Were a parent to ask me how to improve the odds of raising high-achieving children, I would first feel compelled to point out the conclusions of a vast body of well-respected research that is reassuring or devastating, depending on your perspective. The parenting choices that many caring parents agonize over have much less effect on children's outcomes than we have all been led to believe. For all our efforts, "it is very difficult to find any reliable, empirical relation between the small variations in what parents do—the variations that are the focus of parenting—and the resulting adult traits of their children," wrote Alison Gopnik, a prominent developmental psychologist, in her book *The Gardener and the Carpenter*.

So why look at these families at all? To start, their stories are compelling, in that they represent a certain kind of extreme, by design or circumstance—and, yes, at times, they are inspiring. If there are outliers in the data regarding family research, these families would be candidates as intriguing as any. Statistically speaking, parenting probably matters the most in the extremes, at the "tail ends of the distribution," Conley told me. "If you want to make your kid a virtuoso, an all-American athlete, or a chess grand master, you absolutely need to nurture that talent."

Some research does indicate ways that parents with less grandiose aspirations can make important differences in keeping open certain pathways that might help their children succeed, ideas grounded both in research and in the idiosyncratic stories of the families whom I explored. Set high, but not unreasonable, expectations, while offering needed support. Try to unpack for your children the actual, practical value of the kinds of classes that keep professional opportunities open (math and science). If possible, seek out environments that will expose your children to highly educated role models. Consider living in a college town or an area where innovators congregate. And then get out of your children's way. As one mother—someone who raised one wildly successful entrepreneur and two doctors in leadership roles at major hospitals—told me, when I asked her what she thought she'd done right: "I didn't break them."

I started writing this book at a different time in American parenting, which is to say long before the start of the Covid pandemic. The country had its first Black president; we were a country, it seemed, in a state of continuous improvement, in which promise could meet its full potential. The economy was strong, opportunities once closed off were on offer, and well-publicized research was interpreted to mean that all we had to do for our children to succeed and thrive was give them the right growth mindset and figure out how to make them grittier—more persistently and passionately hardworking.

The pandemic was a humbling phenomenon for so many institutions: for hospitals, which were overwhelmed by a virus for which there was initially no vaccine; for government agencies, which struggled to create consistent and helpful messaging; for the press, which sometimes floundered amid conflicting and fast-changing research findings. I would argue that it was also a phenomenon that revealed the limits of parental influence. Schools closed; children suffered; rates of depression, anxiety, and suicidal ideation, already on the rise, accelerated into nothing short of a national crisis. It was impossible to ignore the larger societal forces that were acting on our children, as loving, careful parents functioned, at best, like cairns in a blinding storm: visible, present, guiding, but helpless, clearly, to stop the storm itself. Even after the pandemic ended, my husband and I struggled to find the right balance going forward. We were trying to step back from the unhealthy academic pressures that make so many teenagers miserable, while also nudging them out of the safe havens that Covid-19 made of our homes. So many young people were forced into a defensive crouch at a time when they most need to explore, to take risks—to push the boundaries and to see how far they can push themselves.

As a parent, I have been appropriately humbled by the recognition of how little control I have over my own children's motivations and stances toward reaching goals, for all the insights I've tried to extract from the families I studied. When Leo, one of my twin boys, was about eight, for example, we started giving him piano lessons. As is true for many kids, he played diffidently, maybe for ten or fifteen minutes at a time. Sometimes we urged him to practice more. I know we briefly tried some kind of reward for putting in more time, in the hope that we could just get him to the level at which his competency would allow for deeper pleasures. When that didn't work after a few years—now I think it's laughable that we ever imagined it might—we decided that maybe it wasn't worth it at all to push him. "Let's just stop," we'd tell him. We made it clear that we weren't punishing him—we just didn't see the point if he wasn't interested. And yet he didn't want that, either. So on we went, paying for lessons, eyeing the lonely piano when we passed by it, occasionally nudging him toward practicing—sometimes successfully, mostly not—feeling compelled to keep shelling out money rather than deprive our son of the possibility of cultural enrichment, even if he made it clear that he didn't feel like putting in the work that actual enrichment would require.

Leo turned fourteen in July 2020, a time when Covid-19 was still disrupting much socializing. His friendships, outside of school, had shifted toward a lot of remote gaming, and it turned out that he could not seem to get better at the one video game that those kids seemed to prefer. As he grew bored and frustrated by World of Warcraft, he drifted more and more often toward the piano, which was just a few feet away from the desk and computer he had begun to associate with disappointment.

He was not even taking lessons that summer, which is when he picked up some sheet music sitting on the piano—a Chopin nocturne that I had attempted (inspired, in fact, by a description of it that came up in conversation with Gang Chen, a member of one of this book's families). My son's usual teacher would likely have said it was too challenging for him at that point, but he started working at it every day. For weeks, we all tuned out the repetitive, choppy sounds of someone stumbling and hacking his way through a thicket of notes. And then, one day, as I was making dinner, I suddenly noticed: From all that noise, a melody—haunting, precise, and elegant—had at long last emerged. My husband and I were bowled over, as was Leo: He was as surprised as anyone to find that the more he played, the better it sounded. He played it over and over, until it dawned on him that there would be other, equally beautiful pieces he could master, and so he went on to the next, and then the next, sometimes playing for three or four hours during the course of those long summer days.

This effort was what we had hoped for from Leo all those years. And so, at first, my husband and I were proud and thrilled: Our son was playing piano for several hours a day. And yet, surprisingly soon, that gave way to something else: Our son was playing piano for several hours a day, sometimes, it seemed, all day. Our house is quite small. We both worked from home. The resounding chords, the bumpy hours of practice, the driving pace of a particular rondo that Leo loved, an earworm of a melody that burrowed so deep it sometimes kept me awake at night—this all started to feel like some parable, a fairy tale with a hidden message of warning for parents. My husband and I had furtive conversations in which we pondered whether there was any way to be parents who supported our son's passion while also being parents who asked him to play a little bit . . . less. We gently raised the possibility of an electric piano with headphones (just for the early-morning hours!), which he politely declined. We invested in superb noise-canceling headphones and consoled ourselves: It could have been the drums.

What do we really want for our children? And why do we want it? What is our hope for what will come of it? We want them to be happy. We want them to have choices. We want them to feel the thrill of mastery.

We put the piano in front of Leo. But only he can decide how to play it.

ONE

The Groffs

Memorial Day, the unofficial start of summer, is still sometimes cool in Cooperstown, a village in upstate New York
where snow can cover patches of shaded lawn well into
the first week of May. The gap between the promise of the coming
summer and the reality of the weather conditions in that town was of
some interest, in the eighties, to the Groff children of Cooperstown:
Adam, the oldest, now a doctor and serial entrepreneur; Lauren, the
middle child, now one of the most accomplished novelists of her
generation; and Sarah, the youngest, currently a clinical psychology
student pursuing her PhD, but also an Olympic triathlete–turned–
world-class Ironman competitor.

Family tradition, as Lauren recalled, dictated that on that weekend,
only weeks after the last snows had melted, their father, the esteemed
Dr. Jerry Groff, would roll back the canvas covering of their pool. His
children—a son and two daughters, both tall for their age—would
watch quietly, arms clutched around their chests, wearing their bath-
ing suits, as the tarp was rolled back, revealing all manner of accumulated debris on the water's surface: unidentifiable algae, congealed

muck, a dead frog, some floating earthworms. Some years, a thin coating of ice concealed what was below. Jerry Groff maintains that all of that would have been cleared away by the time any children jumped in, but he does recall the children, shrieking and laughing, teeth chattering, as he timed the event: Anyone who could stay in for a full minute would get a dollar. Lauren experienced it more as a competition; she recalled all three children standing still in stoic patience with fingers practically blue, willing time to pass, each of them hoping to display endurance that would surpass that of the other siblings. Who would bear the cold—who would stay in the longest?

This was a challenge of fortitude, if a low-stakes one; later, Sarah and Lauren would refer, half in jest, to events such as this one as the "feats of strength" exercises that characterize their family activities, even as adults. It was an exercise in toughness for kids who did not have it tough. As a child, their father had known privation, but now he had enough money to provide his family with a rambling, historic house on a lake and a pool besides. A pool might be for lounging in on puffy, airy mattresses, for flirty birthday parties, for relief from the heat, but it could also be an excuse for his children to test their mettle. Competition was fun. Pluck and determination, sometimes mixed with brute force—the family came to call that combination of qualities "Groffiness"—could be fun.

Years later, Sarah would make her living as an endurance athlete. Twice she has been an Olympic triathlete, a sheath of female muscle competing against some of the world's most gifted athletes. In 2023, as the mother of a two-year-old, she won the women's Ironman European Championship in Frankfurt, swimming 2.4 miles, biking 112 miles, and running a marathon, all in eight hours, fifty-four minutes, and fifty-three seconds.

Lauren, too, would rely on endurance to make a living. As a novelist and short-story writer, she does it privately, in a converted bedroom her own two children do not readily access. For five or six uninterrupted hours a day, she builds humans, bits of imagined bone and flesh and feeling, characters who populate universes, whole worlds moving through time. She creates her own reality, in order to reflect something yet unseen about the one in which she lives. She writes a draft longhand, throws it away, writes another draft, throws it away—a dozen or so for *Fates and Furies,* her breakout novel, one of three books for which she has been a finalist for the National Book Award.

Adam, too, would prove an exemplar in a certain kind of endurance: He simultaneously acquired an MD and an MBA, then went on to become a serial entrepreneur in the health-care field. Adam frequently jokes about being the least talented of his siblings, but Lauren credits Adam, the least publicly celebrated sibling, as the one who set the bar for the rest of them. "He's ridiculous," Lauren told me the first time we spoke. "He has always been the star." She still believes that to be true—that within the family, his words carry the most weight.

Jerry and his wife, Jeannine, raised their children in an atmosphere that prized propulsion, motion, effort. The parents had endless projects and focused ambitions for themselves, but their children were never the main one. What the Groff parents expected of their children was primarily a work ethic—labors of love and otherwise.

That the Groff siblings have diversified so remarkably, however, is perhaps a reflection of the general intensity of their home life. Differentiation could serve as a form of refuge from the challenges of coming of age in a family in which so much talent and industry and mental toughness align.

ON FATHER'S DAY IN JUNE 2016, the Groff family assembled in Orford, New Hampshire, at Jerry and Jeannine's then-new home, about twenty-five minutes from the Dartmouth Hitchcock Medical Center, where Jerry, a rheumatologist, had completed a two-year fellowship decades earlier and where Adam also works. The home, like the one in Cooperstown in which the children were raised, is sprawling, with grounds that are pastorally impressive: rolling lawn, a barn, a murky but swimmable pond, ringed by willows and trees. Inside, Jeannine has painted the rooms in historically appropriate colors, seemingly never-ending work that she proudly points out to guests; they can afford to hire people to help, but the Groffs still prefer to take care of that kind of work themselves. Parked by one of the outbuildings is usually a pickup truck that Jerry drives around town.

Jerry and Jeannine had recently moved from Cooperstown to be closer to the families of two of their three children. Sarah, the youngest, lives nearby with her husband, the professional distance runner Ben True. Adam, the oldest, lives in that area as well, with his wife, Tricia, a pediatrician, and their four children. Lauren, married to Clay Kallman, a property developer with whom she has two children, frequently spends her summer there and eventually rode out the Covid lockdown in another home on the property in Orford, a modern space where she wrote while her two sons were taking classes online.

Sarah and Ben arrived at the gathering a bit late, on their bikes and wearing athletic gear. Although Sarah can be playful and lively in the many interviews she does on podcasts, that day she and her husband were both quiet, radiating a perceptibly tense energy that even I, a guest at the Groff home, could feel. The upcoming weeks were potentially momentous for both of them, and their minds were drifting forward to how they'd both fare physically and mentally in competition. Ben, who still holds the American record in the five-kilometer road race, was trying to qualify for the 2016 Olympics in the five thousand meters, having narrowly missed that opportunity four years earlier, hampered by a bout of Lyme disease. Sarah had already qualified, but that only intensified her stress, as, at age thirty-four, she was likely facing her last chance at a medal. She had spent the four years leading up to this moment training in what her family calls a "bubble," a zone from which she rarely strayed, missing friends' weddings or nieces' and nephews' birthdays, pushing herself through workouts so grueling that they left her mind drained of thought. She had run in hundred-degree, high-humidity conditions in Mexico and swum in fifty-degree waters in the Pacific, had woken up multiple times in an ambulance after passing out and once, for a two-month stretch, trained in excruciating pain with what turned out to be a fractured sacrum.

Sarah had met Ben not long after relocating to New Hampshire, where she had moved to find some stability and to be closer to Adam and his family. Adam is not quite so long of limb as his sisters, and yet he is a serious runner as well, a marathoner who will occasionally join Sarah on a long run. He had even done so recently, Tricia mentioned, as the family sat around a table eating lunch.

"Oh yeah? How'd that go?" Lauren asked, a little amused. There was an edge to the question: She seemed to like the idea of her older brother struggling to keep up with her younger sister.

Adam's daughter Heidi, then four, was a blond bit of a girl, her dress pink, her cheeks pink; she also wore a pink cast on one leg, the result of a mishap with a sprinkler just a few days earlier. On the cast, Jeannine had written, "You are so tough!" As the family lingered over their meal on the patio, the foothills of the White Mountains visible in the distance, Heidi, who was without her crutches, started off toward the kitchen, where she had heard there were live lobsters. Various assembled members of the Groff family watched as Heidi dragged herself, in a cross between a military crawl and the toddler scrabble she'd so thoroughly outgrown, across the patio toward the steps leading up to the back entrance of the house. When she finally arrived at the steps, she looked up balefully. The question hung in the air: Was she going to keep going? Sitting there watching, well aware that I was among trained medical professionals, I worried nonetheless: I pictured tiny bones rebreaking, the cast splitting open. "Let her do it," Tricia said. "Let her see how far she can get."

There was no fanfare when she arrived at the top, but her grand-mother repeated the ultimate Groff compliment: "You are so tough!"

The Groffs almost never talk talent; instead, they talk a lot about toughness, a family value that is more about resourcefulness and self-sufficiency than it is about beating the competition.

Not long after lunch, several members of the Groff family went for a swim in the small pond on the property, with Ingrid, Adam's older daughter, then six, clinging to Sarah's back as if she were riding a very careful, slow dolphin, the lily pads parting in their midst. Everyone but Lauren, who was dressed in black, went in; this was a family of game parents and game children, and anyway, it was hot. Finally, everyone trudged out to make the five-minute walk back to the cool main house. Ingrid was about halfway there when she collided with the family's dog and suddenly started to cry. "You are fine, sweetie," Tricia reassured her. Ingrid had been complaining of a number of ailments since her sister got her cast. When Adam had joked, at one point, "No one else is allowed to break their leg around here," Ingrid had burst into tears. "I'm not even allowed to break my leg?" she had sobbed. I wondered if what she really craved was the chance to prove herself brave, even at the cost of significant injury.

It took a few minutes for Tricia—herself an ultramarathon runner—to realize that Ingrid's foot was bleeding, that the full weight of dog had landed on it. She comforted her sniffling daughter and cleaned and bandaged the child's bleeding foot. Tricia briefly felt terrible; but in the long run, Ingrid would be no worse for wear. She had even found a way to prove herself brave.

THE GROFF HOME IN COOPERSTOWN, New York, had some of the gracious touches one would expect in a late eighteenth-century colonial, but it also had acquired, long before Jerry and Jeannine purchased it, an additional wing with a seventies-style interior design that kept the whole enterprise feeling less than grand, despite its con-siderable size. It was not a home for design purists, but in square foot-age, it was a visual testament to its own upward mobility.

Jerry Groff poured work—his own labor—into the house, with re-sults that were a source of pride for someone who was an overcomer in his own right. The Groff parents' careers may not have been as public as their children's, but the life they built for themselves and their children is a triumph in and of itself, given the hardships of the homes in which each parent was raised.

Jerry grew up in rural Pennsylvania, the child of parents who strug-gled financially and as a couple. The circumstances were dire enough that Jerry and his three brothers all spent a brief period of time in a boys' home—an orphanage—while their parents tried to resolve their differences. They ultimately divorced when Jerry was seven, at which point his mother's job working the night shift at a diner became the family's main source of income. His mother was exhausted, his father, a spotty presence; Jerry and his brothers all but raised themselves. Jerry, number three in the birth order, somehow coasted along, per-haps free of the burden an oldest sibling might feel or the vulnerabil-ity of the youngest. He was, insistently, a sunny optimist; he was also a natural athlete. A grandparent's generosity ensured that he attended a strong school starting in seventh grade—one of the first Waldorf schools in the country—which set him on an academic path that landed him a scholarship at Franklin & Marshall.

Once there, Jerry recognized a kindred spirit, an appealing young woman who joined him as a member of the school's first coeduca-tional class. Like Jerry, she was shedding a hardscrabble past and in-venting a fresh version of herself on a college campus. That young woman, Jeannine, sometimes felt, looking back, that she had not just left home, but she had escaped with her life, an outcome of which she had not always been certain throughout her girlhood. Her father drank, which made him unpredictable and on rare occasions violent, a man who fought mean and with little restraint. When he was on one of his rampages, it fell to Jeannine, the oldest of four children, to hustle her siblings to the attic of the barn, a dark, tucked-away space with no electricity or heat; sometimes in the cold, usually in the dark, they'd stay there, waiting out his rage until she made the judgment it was safe to return to their home.

Jeannine was the only one of the four siblings whom her father never hit—she was regarded as "special," a burden that brought its own anxieties, which she channeled into a furious focus on what she could control. She started twirling a baton as a young girl, a diversion that became more than just a fun way to pass the time; she attacked that hobby with hours and hours of self-driven practice, escaping, wholly absorbed, into the skills she was building. The hours paid off and offered her an out from the chaos of her upbringing: She emerged a competitive twirler, a pageant girl, and a virtuoso performer in sequins and go-go boots. In high school, she worked over the summer at a biochemistry lab but was also in charge of choreographing hundreds of dancers and baton twirlers at state football halftime shows; she could juggle three flaming batons, a skill that served her well when she entered the Miss American Teenager pageant. (She was a finalist for the crown.)

If Jerry was an amiable, stalwart worker, Jeannine was a dazzler, the college's head majorette as a freshman, with a scholarship of her own. At the time, athleticism in traditional sports was a skill with little social capital for girls—but baton twirlers were embraced, and Jeannine was a champion, the winner, as a local paper made note in 1969, of "25 trophies, 15 medals, and several national and state titles for baton twirling."

In high school, Jeannine won prize money for a science scholarship, and she intended to channel her drive into getting a medical degree after college, even though, at the time, women made up less than 10 percent of medical school students. After she received a few poor grades in her pre-med classes at Franklin & Marshall, she called home, discouraged. "I told my parents I was thinking about quitting, and they sort of said, 'Yeah, that sounds like a good idea,'" she told me. And so she quit the pre-med track and majored in biology. Instead of going on to medical school, Jeannine worked in a lab so that Jerry, by then her husband, could get his medical degree, an investment that landed them, eventually, in their showy white lakeside home in Cooperstown. Jeannine's decision to curtail her ambitions would trigger an ambient, restless regret for many years to come, one that became its own quiet force in the family she'd eventually form.

In short order, the couple had three children. Adam was active, curious, confident, constantly in motion. Lauren, born eighteen months after Adam, was quieter, more still. She differed so remarkably from her brother that her parents worried that she was not developing normally. They brought her to the pediatrician for an evaluation, only to be told what parents find so hard to believe: that two children born to the same parents could vary temperamentally so much—their differences manifest even in infancy—that having the same family environment would not elicit the same kind of behavior. "All parents are environmentalists until they have their second child," wrote the social psychologist Marvin Zuckerman. By environmentalist, he meant a person who believes that a child's environment, including how they are parented, is far more influential in their development than their innate, genetically influenced tendencies.

Lauren developed into an intense girl, a parent pleaser and a dreamer and, to the annoyance of her highly rational brother, a storyteller. The two, she has written in her recollections, "brawled to blood and tears," with Adam teasing Lauren mercilessly. When Lauren was three, her mother gave birth to Sarah, a girl who'd love birds and art; her relationships with her siblings were gentler, her stance toward them more deferential than provocative.

All three children were athletic, and through the family's relationship to sports, guests at the home could catch a glimpse of how the Groff parents felt about competition. "You'll never hear an 'Oh, that's good enough,' from the Groff parents," said Pat Murray, Adam's oldest friend, who was a steady guest and somewhat fascinated observer. "But now that I think about it, that was not pushy somehow. If one of their kids accomplished something, it was almost like 'Okay, great—what's next?'"

Lauren, independently, would use that same language to describe her parents' attitude toward success in an interview she did with her sister during the run-up to the Olympics: Their usual response, she said, was "What's next?" "There was always a feeling that whatever we did was not quite good enough," Lauren said. "We always had the potential to do better." That orientation toward the future, said Lauren, meant you always had room to improve; it was both the whip and the balm. It was the reminder never to settle for good enough and the reassurance that whatever happened today could be surpassed, with a little effort, tomorrow. "Do your best and forget the rest," Jerry Groff also used to say if one of them flubbed a pass in a soccer game or lost focus in a race, advice that Lauren found comforting but that could also be maddening in its deceptive simplicity. What if you hadn't done your best? ("Crushing shame," said Lauren.)

The Groff parents set their expectations high, but they were not the type to watch over their children while they practiced an instrument or read their papers before they turned them in or harped on them about college applications. They didn't come to practice; they didn't hover or coach the coach.

"They were there for us when we walked up to them," said Lauren, "but otherwise they let us do our thing." Or, as Adam put it, "They had no idea what we were up to most of the time."

The one ruling principle was the expectation of exertion—get off the couch, get to a game or a practice, give it your all. In the Groff home, Adam's father even chastised him, when he was a teenager, for spending too much time on a slow desktop computer the family had acquired. "It was much better to be in the real world than on a computer," Adam said his father believed. By then, the early nineties, kids who had whiled away their childhood hours poking around early programming languages were already making millions in the dot-com boom. But at the time, it looked to Jerry Groff like a pointless, slightly unwholesome time suck—not "Groffy."

"You know how sometimes you just want to kind of zone out on the couch and watch TV and hang out?" said Lisa Trever, a friend of Lauren's, now an art history and archaeology professor at Columbia. "You didn't do that at the Groffs'." In the Groff household, she said, there was a sense of urgency imbued even in leisure. "There was this bright, burning energy that suffused the entire house," recalled Lisa. "They pushed themselves very hard—very."

For the most part, Jerry and Jeannine let their children function independently because they were too busy to have any other option. A science teacher, Jeannine also poured considerable energy into their large home, forever windexing acres of plate glass or repainting the interior, entering road races—and winning—in her age group when she took up running. Jerry poured hours into long-distance running when he trained for marathons; on the weekends, he worked on building a second staircase in the house, a project that took him five years to complete as he learned on the go.

"Work was holy," Lauren said.

"If parents are the fixed stars in the child's universe, the vaguely understood, distant but constant celestial spheres, siblings are the dazzling, sometimes scorching comets whizzing nearby," the developmental psychologist Alison Gopnik once wrote. All of the Groff siblings speak, with some reverence, of their parents' industriousness and achievements. But they each claim that the siblings shaped one another as much as their parents did, sharpening some of their edges, smoothing others, forcing one another into shapes that could coexist within the space of their family.

Lauren Groff has a memory of playing in the snow. First the Groff children were building snowmen, but then they were building snowmen to see whose could be the biggest, furiously rolling balls and holding their breath as they set one massive white boulder on top of another, dreading its collapse. "It was all very lovely, until suddenly it flipped," said Lauren. "And then I was crying—I was always the one crying the most." She recalled feeling the constant effort of self-assertion: "There was this constant jostling." She now looks at Adam through the eyes of a mother of two boys: Perhaps he was the tormentor of his little sister because she was not so little. She upset the natural order of things with her height. Adam was preoccupied with his own distractions, an adolescent largely oblivious to how exquisitely sensitive his sister was, how seriously she would wrestle with what she considered his low assessment of her. As for Adam, he barely remembers any heat of emotion between them, beyond the intermittent squabbles any two children close in age would have.

"I was an immense introvert with an older brother who wouldn't let me speak," Lauren Groff once told a reporter from *The Guardian,* "so I was the biggest reader." She generally retreated into her room, into the books that kept her company there and provided emotional refuge. "Without it"—by *it* she means Adam's older-brother teasing— "I never would have read six billion books," she told me. In *The Monsters of Templeton,* her most autobiographical work, set in a town much like Cooperstown, she wrote, "When I was small and easily wounded, books were my carapace. If I were recalled to my hurts in the middle of a book, they somehow mattered less. My corporeal life was slight; the dazzling one in my head was what really mattered. Returning to books was coming home."

Adam is now an adult she admires: a caring father and brother whose work entails developing more humane kinds of health care. But when she was a child, the more Adam teased her, the more fury burned in her, the less genteel and submissive she became. "The mere fact of his being older and smarter shaped me in particular, so as a young woman I was just trying to keep up and trying to show him that not only was I a worthwhile human being, but I was equal to him. This was the seventies and eighties, when feminism was becoming much louder," she said. They were fourteen and twelve when Adam made one joke too many at her expense, and Lauren—a strong soccer player—kicked him in the groin with all her strength. ("I *do* remember that," Adam said.) That kick became symbolic of a specific kind of resistance that would fuel both her creativity and her drive—even the content of some of her books. Eighty percent of what drove her to push herself in life for many years, Lauren estimates, was Adam.

Among the siblings, there was a trickle-down effect of competition, one that seemed to increase in concentration, as if passing through a filter. Sarah was competitive, from the outset, but only with Lauren, not Adam. Sarah's first interest in swimming competitively, she believes, was inspired less by a love of the sport than by her desire to surpass her sister after Lauren, at age eight, joined a local team. "She told me that no way was I allowed to join the swim team," said Sarah in an NBC video in the run-up to the 2016 Olympics. "I really had minimal interest in swimming, but of course at that point my goal was to beat her in swimming. Not just join the team but beat her." (Lauren, to this day, insists that she was merely the messenger—that according to the rules, Sarah was too young.) Whereas Lauren's desire to show Adam what she could do was driven by a simmering state of righteousness, Sarah's came from a place of adulation: "I wanted to beat Lauren," she told me, "but I wanted to be her, too." To Sarah, Lauren was a role model, the girl who taught her how to be a girl, to love books, to tie a shoelace—to *believe* she could tie a shoelace. At age eight, in 1986, Lauren had somehow locked onto the words "You can get it if you really want," the title of a Jimmy Cliff tune that she crooned to Sarah in the back of the car on some family trip, as she showed her patiently how to make a loop with the laces of her sneaker, over and over again.

If you are going to become an Olympic triathlete, it helps to have grown up with an encouraging sister. It also helps, in a way that maybe also hurts, to have two older siblings who have left their mark on a small-town school before you are old enough to leave your own. "I didn't feel like I was trying to measure up to my siblings to please my parents," said Sarah. "It was for me—it was omnipresent. I was never going to be good enough, and that was a failure of something about me—because we all had the same upbringing. I don't think that I would be a professional athlete if it weren't for my siblings. I wouldn't have tried to carve out my own little world the way that I have."

Sarah's experience as a younger sibling may be common among serious athletes. By 2014, 257 female soccer players, each of whom trained for the US women's national team in their age category (ranging from under fourteen to twenty-three), had filled out a survey that asked general questions about their lives. Close to three-quarters of the girls and young women playing at that elite level, the researchers found, were younger siblings. They were forced to play up a level if they wanted to hold their own with their older siblings, theorized April Heinrichs, development director for US women's soccer (and a former national player herself) and Matt Robinson, head of the sports management program at the University of Delaware, both of whom conducted the study. The constant competition may have improved their game, relative to their peers, though it was also possible that their parents could better navigate the system, having been through it at least once and knowing more about the sport. That younger child is also likely to have started earlier than an older sibling. It might not dawn on a parent to hand a two-year-old a soccer ball, but if her five-year-old sister happens to be playing with one, the younger one might reach for it.

Jonah Berger, a professor of marketing at Wharton, has a different interpretation, one that rings true to Sarah: Younger siblings focus on sports, more than their older siblings do, because that is the niche available to them—the domain where they can stand out. Consistent research has found that oldest children tend to do better academically than their younger siblings, perhaps because they are the only siblings in those formative early years and have their parents' undivided attention. Certainly, in the Groff family, Adam was the academic star; Lauren, too, was a bookish natural student.

Lauren was also a superb swimmer, good enough that she competed in the Junior Olympics. But even still, as a choice, running and swimming felt "safe" to Sarah, as she put it. "They weren't going to touch it," Sarah said. "I saw that Lauren didn't have the same interest in being an athlete that I did. She didn't have that desire." If Sarah put in enough work, she thought, she could beat Lauren. The lake outside their front door gave her one opportunity.

Otsego Lake: It was as much a part of their home as the house, and swimming the length of it loomed as a perennial dare for Sarah. It was also, for the Groffs, one of the factors, along with their pool, that might have given them a swimmer's edge. Wayne Gretzky grew up with an ice hockey rink that his father built in their backyard; Jimmy Connors had a tennis court in his backyard in working-class East St. Louis; and Mickey Cochran, a passionate skier, built a functioning ski slope in his own backyard, with a four-hundred-foot rope tow, on which he trained all four of his children, each of whom became Olympic competitors. Ease of access alone won't make a champion, but having a tennis court in one's backyard, or a ski slope or a pool, can reinforce whatever innate connection to a sport an athlete already has while removing a great many obstacles. In addition to enjoying proximity to their pool and the lake, the Groffs were within walking distance of an athletic facility with outstanding equipment, trained professionals, and affordable after-school sports programming that was the pride of Cooperstown.

When Sarah was fourteen, in 1996, she told her family she wanted to swim the nine-mile lake, as much as an eight-hour proposition for some recreational swimmers. Her parents heard her announcement as a remote possibility to be scheduled at some distant time in the future. "Honey, that's a long way," Jerry told her. But Sarah was somehow confident she could do it. "I probably undertrained," Sarah says now. "I knew other people had done it. I could swim."

Just two weeks after she made that announcement, Sarah said she would do it that Sunday morning.

Her father had a moment of trepidation. "For one thing we had breakfast plans," he said.

But holding him back was another, more emotional and nagging thought: "I think foremost was trying to prevent her from failing—there was no logic that said she could do this." He and Jeannine did not want her to have to give up, to experience that disappointment.

Jerry looked at his daughter's determined face and realized that he would need to rise to the occasion. "This is going to be a disaster," he told his brother who was visiting. Then Jerry asked his brother and sister-in-law to drive alongside the lake with Jeannine in case Sarah needed to cut the swim short.

The day of the swim, Adam joined Jerry in the family's aluminum rowboat to accompany her and time the swim. With Sarah, who was five years his junior, Adam felt none of the chafing he and Lauren experienced with each other as children; if anything, he felt protective. Three hours and forty-nine minutes after they started out, Sarah stood, shaking a little, in the muddy sand at the Springfield shore. Adam's hands were bloody with blisters. Sarah had not just swum the whole nine miles but had also broken the town record for female swimmers of any age.

"It was a way of distinguishing herself," Lauren Groff said. "But it was also a way of conquering this place. She was a third child—it was psychically huge in her life. It turned her into who she is, and not just as an athlete. It showed her who she was."

For Jerry and Jeannine Groff, that day meant something else, something that would benefit her siblings as much as it helped Sarah. "That helped educate us about our kids' capacity," Jerry said—a reinforcement of a Groffy sentiment that unless they let their children risk failure, they'd never find out just how far they could go.

About a year after Sarah completed the swim down the length of the lake, Lauren decided she was going to do it, too. When she was about a mile from finishing, her father, who was canoeing alongside her, looked at his watch and realized that if Lauren picked up her pace even a little, she was going to beat Sarah's time. He thought about it for a moment. Should he mention it? The thought of her beating Sarah's record time pained him. But why would he hold Lauren back?

How to handle competition among their children is, for many parents, an excruciating and recurring struggle: Love may not be a pie, to paraphrase the title of an Amy Bloom short story, but, in sports, there is only one slice for first place. The parents of the identical twin Bryan brothers—Mike and Bob, considered by many measures the most winning doubles tennis team in history—addressed the problem early on with an extreme solution.

The boys' mother, Kathy, ranked as high as number eleven nationally on the women's circuit in her youth, and their father, Wayne, played in college. Together, they were part-owners of a tennis club where both taught for years. The Bryan parents happened to be experts in a sport that had the perfect outlet for twins: doubles. On a soccer team, one boy might score more than the other; in track, one might be faster. But on a doubles team, their interests were perfectly aligned.

From the beginning, the Bryans had a "master plan" for their twins to become tennis champions, Wayne told *The New York Times* in 2020: no television or video games, ever; early exposure to imaginative, kid-friendly games involving a racket and a ball; and no direct competition between the two boys. When one brother, as a boy playing around, hit a ball over a tall tree in the family's yard, Wayne spent three days making sure that the other one could accomplish the same feat. His intent was to maintain confidence: How could an athlete win on the court if even at home he felt like second best? There was already enough natural competition between the boys, he believed, that he should avoid stoking it further.

To the reported frustration of the tennis leagues in which they played, if both boys made it to the finals of a tournament, they would default and share the trophies for winner and runner-up. That kind of intentional leveling sent two messages, the first of which had only a slim possibility of being true: Each was the other's equal in every way. The second message was undeniably true: The parents wanted winners (when they played doubles), but they wanted no winners in competition of the boys against each other.

Joe Kennedy, father to John and Robert Kennedy, was known to say, "We don't want any losers . . . in this family, we want winners!" The proclamation was likely painful in a family whose members were forever competing with one another in tennis and touch football and sailing. Did it mean that whoever came in second, in all those competitions, was unwanted?

The Bryan boys went on to dominate in doubles until, inevitably, fate weighed in where their parents could not: After winning at least 1,100 matches together and winning more male doubles Grand Slams than any duo in history, one of the twins—Bob, who was born two minutes after his brother—injured a hip in a match in Madrid and had to bow out of competition. He underwent hip-resurfacing surgery and watched, walking with a cane, as his brother trained for Wimbledon with another player, forming a new duo that went on to win that Grand Slam, as well as the US Open, victories that placed Mike first in the record books. By the spring of 2019, however, months after the surgery, Bob was back, winning with his brother the Miami Open, in a comeback that sports followers consider one of the most unlikely and remarkable in tennis history. The brothers finally retired a year later, at age forty-one, finishing a career that itself breaks records for endurance.

The Groff sisters ran track on the same team in high school, when Lauren was a senior and Sarah a freshman. The coach never let the girls compete in the same race, which Sarah thinks was an act of not just kindness but also wisdom. "We would have been each other's best competition," said Sarah, "but it would have changed the dynamics of the relationship." Competing against each other, she said, "might have made us incredible athletes, or destroyed us."

That day on the lake with Lauren, Jerry decided his children would have to handle whatever came of that competition. "Just so you know," he told Lauren as neutrally as possible, "if you picked it up a little bit, you could beat Sarah."

Lauren remembers hearing her father tell her that, midswim, the fog still hovering above the water, the sound of the oars slicing the surface. "To be perfectly frank," she told me, "I'm a very competitive person, so there was this war going on. There was a very complicated, loving, familiar conversation going on in my head. For sure I wanted to beat her time. But I think he said it just the right way. 'If you want to—but you don't have to' was implied." Also implied, she said: "I want you to think about this pretty deeply when you're swimming alone in the middle of a giant, cold lake." The way he phrased it, Lauren thinks, clearly showed what he thought was right. She kept swimming at precisely the same pace. "Sarah was young," she said. "It was the thing that she was really proud about." *You can get it if you really want.*

Sarah, hearing about this moment for the first time many years later, had her doubts about whether her dad was even right about the timing. But for Lauren, at any rate, that day at the lake is a fond memory, not a missed opportunity. "I remember cruising into Springfield," she said. "I could stand up in the muck, but I couldn't figure out how to get to the docks. I'm very blind." The whole time she was swimming, she said, she was wholly dependent on her father. "I just had to trust that he was leading me in the right direction."

IN THE GROFF HOUSEHOLD, "we were not allowed to fail," said Lauren. What was failure? "Getting a B," Lauren said. It wasn't that anyone got angry. "It was just, it would make my dad sad. You didn't want sad Dad. You wanted Dad to be proud." Jerry said he doesn't recall reacting strongly to any of their grades—but does concede that his expectations were high, expectations that came to Sarah indirectly, via her siblings. "Lauren and Adam didn't get Bs, so I have to get As," she said. "That's just what Groffs do." Jeannine was more likely than Jerry to lay down rules, to set the tone for a house where loafing, like most sugary foods, was off limits.

They could not fail—which was not the same thing as being told explicitly that they had to excel, that they had to be superb. Jeannine, who was overcome with gratitude about the life she was able to lead, about the home she could create for her children, did not actually need them to achieve at any particular level. But her children say they nonetheless picked up on some vague sense of imperative. They saw it in the example she set daily: in the energy she applied to the house, the overhauling of bedrooms with matching bedding and curtains all completed at her hands, the frustration she expressed privately, at home, toward students she deemed mediocre. (Lauren said she was always told that her mother, as a teacher, was "strict and very exacting," even a little intimidating.)

Tricia Groff, Adam's wife, met Jeannine when her future mother-in-law was already well into middle age, and was struck by how demanding she was of herself. "She's relentless. 'Don't give up. Perfection. It has to be just so,'" said Tricia. At age fifty-two, when many of her colleagues were thinking about retiring, Jeannine decided to go back to school to be a physician assistant, commuting an hour and a half each way to classes in Albany and receiving a master's degree in 2007.

Looking back, Lauren recalled a household imbued with anxiety. "It was a deep and wordless energy—an electric current that sets fire to your own propulsion," she said. "It's a gift with spikes." The home Jeannine created for her family was safe, sparkling, gracious; the food was healthy, the routine, orderly. But Lauren was always aware that something powerful drove her mother to make it so. "A lot of her hunger comes from deliberately turning her back on this darkness and facing the other way and wondering why there's a strange feeling at her back the whole time," Lauren said.

Jeannine rarely talked to her children about the sporadic brutality that left so much uneasiness in its wake in her childhood home, but she said she'd reached some understanding of how it had shaped her. "Violence—that is the root of it," she said. "It makes me comfortable if I can control my environment."

The Groff children have a hard time putting their finger on how they knew that there was something defensive in all that productivity—that Jeannine had something to prove. "There's half of her that says she's worth it, that she's smart, and that she's all these things—that she's something. And then there's the other half that fears she's nothing," said Sarah. She has thought about that quality in her mother a lot, because she believes she has that same divide; she said she's also seen it in Lauren. Even Adam—whose trademark trait might be confidence—feels some version of that tug. "None of us ever stops working," he said. "What the three of us have in common is the idea that whatever we do, we always feel like it's not good enough." Tricia sees perpetual dissatisfaction as perhaps the most essential trait of the Groffs. "They've all done amazing things, and they still tend to feel like it's not enough. If you were to say, 'Was that good enough? You're all set now, it's over?' No, not good enough. Could have been better."

That discontent is part of the drive that has made them successful. Adam helped build a home health service that allows more people to receive acute care in their home and another business that partners with community centers, like homeless shelters and churches, to bring medical care and substance abuse treatment directly to those sites. He also started a nonprofit that allows more people to experience hospice in their own homes. But there is always a next project, a bigger idea, one that he might launch at a bigger scale for maximum impact. "Launching a start-up—it's grueling," he said. "It's really terrifying moments and pressure, and it's stress. It's constant problem-solving." The fear of not getting the funding right, of having to fire people, of failing in the solution he's convinced himself is better than all the other systems out there, is mentally exhausting at times. "If there's one thing I've failed at, it is becoming a good sleeper (and I've tried every tactic)," he wrote to me in an email. "Turning off my brain can be hard, and I think this likely has a cost on my overall health." Describing the Groffs, he said: "We're not content with the way the world is, and where we are as people." His tendency toward forward propulsion is such that he knows he has to be conscious of not letting one business suffer when his attention drifts inexorably toward the next.

Lauren, too, is ambitious: She does not just want to write, she wants to write big, culture-shifting books, books that are wholly original, that will elevate and chip away at people's ideas of what women writers can do. Maybe, she thinks, she is ambitious because Adam is ambitious. Because he entered the world as a white male with resources, he was right to assume the world was his for the taking and the shaping. If that was true for him, why not for her?

When she married, Lauren did not see herself as the kind of author who steals time to write when the baby blesses the house with a nap or in the early hours before she leaves for work (as Toni Morrison famously did). She would grant herself permission to take herself seriously as a professional writer from the outset, to take precautions to ensure that her art would not be subsumed by the traditional obligations she saw looming in the distance. When Lauren agreed, at her husband's request, to move to Florida, where his family's business is located, she had him sign a formal contract asserting that he, not she, would be the primary caregiver, the one who was responsible for getting the children out the door with their lunches. Morning was sacred writing time, and it needed formal protection. When the opportunity came up for her to study at Harvard for a semester on a fellowship when her sons were in grade school, she seized it, traveling frequently back to Florida but confident that her children would weather the separation.

Her work reflects that seriousness of purpose: Her novel *Fates and Furies* was, *The New Yorker* wrote, "hailed for its formal innovation"; *The New York Times* called it "an unabashedly ambitious novel." With her subsequent novel, *Matrix,* Groff turned her imagination and research to the interior life of a brilliant abbess living in the twelfth century, a nun who has vision—ambitions, but also visions. Her most recent novel, *The Vaster Wilds,* follows an unnamed girl escaping Jamestown, Virginia, a vulnerable figure who experiences both the ecstasy and the cruelty of the natural world, all the while in a desperate flight that reveals the almost supernatural physical strength that even a vulnerable young woman can summon. ("For my sister, Sarah," reads the dedication page.)

Lauren's ambition fuels her, but it also nibbles away at her peace of mind. It costs to care that much. There are nights when her anxiety runs away with her sense of self, when she lies awake, miserable, her heart pounding, sleep impossible, as she wrestles with the fear that her work is only mediocre, that there are holes at the heart of her story-telling, that she is not working hard enough. Her husband, Clay Kallman, is often relied on to live up to the promise of his name: Calm Man. A six-foot-six mass of solid certainty, a man who is cool like marble, he does as much talking as he thinks might help and no more; he mostly listens and waits it out.

When Lauren talks about her writing process—the drafts she writes in longhand, then throws away, only to start fresh again on the same book, going through revision after revision after revision—she sounds like someone with extraordinary discipline, a person whose creative process is both brutally demanding and somehow pure. But, for Lauren, who struggles with chronic obsessive-compulsive disorder, that process is just as much a work-around—if she didn't consider every draft fluid, she told me, she would get stuck on any one sentence for six months "and grow to hate myself." Writing, she said, is a way of managing her considerable anxiety. "I think sometimes we write because it's easier than living," she said. Her way of writing allows her to cope, in other words, with her own challenges, without exactly absolving her of the torment that the work also entails—the days of self-doubt, the fear of failing, the loneliness and responsibility of holding together a world of her own creation.

Sarah told me she would never have thought to become an Olympic athlete if she had not seen her sister decide, shortly after graduating from college, that she was going to try to be a novelist. Neither choice was practical; both required faith bordering on arrogance, as well as unusual sacrifice. But Lauren had forged ahead and made her way, without doubt or second-guessing from her parents; maybe Sarah could do the same and succeed.

SARAH ATTENDED MIDDLEBURY COLLEGE, where she first started competing in triathlons and winning. She wondered how far her talent and training would take her and, after she graduated in 2004, her father drove her out to Boulder, Colorado, so she could find out. The unspoken thought he carried with him was the same one he had kept to himself when he learned that Lauren was majoring in medieval French: How is she going to make a living? But he kept driving west and assured his daughter that if she needed a financial cushion, she had one.

Sarah started winning prize money almost right away, and soon sponsors were paying her to wear their athletic products. But even if it's not used, the existence of a cushion is a powerful thing. Having financial security, for someone aspiring to be a professional triathlete, is akin to having a naturally lean, structurally sound physique, an advantage that has nothing to do with grit or effort. It is even, quite possibly, the most consequential advantage that the Groffs bestowed on their children. To the extent that the Groff children took risks, Adam said, it's because their parents, who came so far without the benefit of a strong safety net, gave them the luxury of knowing that, as he put it, "We couldn't fall too far."

Lauren, too, had that cushion as she headed off to graduate school in fiction writing at the University of Wisconsin at Madison. For three years after graduating from Amherst College, she had worked as a mixologist and an events planner, writing furiously during her downtime, free of college debt or the nagging fear of hitting any kind of financial rock bottom. By the time she applied to graduate school, she was engaged to Clay, a man with a family business that would support them both comfortably.

What Sarah thought might be a post-college lark before heading off to law school became a sport she could not leave behind, and she kept moving up through the ranks until she was one of the top competitors in the field. In 2010, when she was twenty-eight, however, she suffered an injury, breaking her sacrum. She trained through the pain, then broke it again. She had been living in Australia in the winter and in Switzerland in the summer, far from any family and friends. Finally, at the end of the year, her parents, who found it difficult to see her in so much physical pain, had a conversation with her, asking if she thought it might be time to stop: She seemed miserable and was depleting her savings. Her parents asked her to think about the past year, and what it told her about what was likely to come. She was not ready to quit, she told them.

She was right to feel her best was ahead of her—the next year, she qualified for the Olympics in London. She raced better than she expected the day of her 2012 Olympic debut—so well, in fact, that she came close to medaling. The triumph of competing in the Olympics merged with a painful sense of regret: She had lost, she realized within minutes of the end of the race, the sweat not yet dry, because she had not believed she could win; she had not believed she belonged with the women who won. Somehow, all along, they were the ones she had been picturing with medals around their necks, not herself. When they had surged at the end, she had hesitated, missing the bronze by ten life-changing seconds.

By then, she was living with her future husband Ben True, a native of New Hampshire who was, even among elite runners, considered exceptionally tough, someone who preferred to train alone in the Granite State rather than join the community of his peers in Eugene, Oregon, or Boulder, Colorado. "Now a True, but still very Groffy," Sarah's self-description on X read. Her life was bound up with his, and their togetherness was in their commitment to world-class athleticism. She wanted the experience of not just competing in the Olympics once again—four years after she'd come in fourth—but competing in the Olympics along with her husband, Ben, whose personal best in the five-thousand-meter race was at the time the ninth-fastest in US history.

The buildup to the 2016 Olympics in Rio four years later required all-absorbing commitment and peak mental and physical endurance, with excruciating emotional upheaval in the last weeks: When the time came to qualify, Sarah made the US team, but Ben did not, missing qualifying for the five thousand meters by less than half a second. An NBC camera caught Sarah, in shock, sobbing, after they found out. She would be going to the Olympic Village alone; she would, as she saw it, have to win for both of them. "She was one of the favorites to medal at Rio," Lauren later wrote in an essay in *The Sewanee Review.* "We all believed she would."

The story of great athleticism, and the American success story, is one in which extraordinary feats of strength, of character, of sacrifice, always deliver. It was a story Lauren believed for her sister: Surely Sarah would be rewarded for all her work and suffering.

And yet Adam, a man of science, looked at the numbers—Sarah's times—and felt less confident. Her most recent times put her in a good position for the top ten; but a lot would have to go just right—some bit of luck on her part, some very bad luck on the part of her chief competitors—for her to finish in the top three, especially given how she raced in high heat and humidity. Even still, his role, he knew, was not to present charts and numbers but to cheer from a close but not too-close distance.

On August 20, 2016, the day of the race, Jeannine and Lauren showed up three hours early, situating themselves in the stands near the finish line. Jeannine did Zumba to the blaring Brazilian music to keep herself busy; Lauren, whose anxiety needed an outlet, jumped up and down. From where they sat, they could see Sarah, along with the other athletes, lined up on a gangplank along the beach.

In a clip on NBC, Sarah once described what it's like to be a competitive triathlete, hearing the thrumming music, often drums, before a race begins. "I think like *Lord of the Rings,* like the orcs are going to come and eat me," she said. "It's absolutely terrifying." She always knew how much could go wrong—and how much pain the race would entail, even if all went right.

By the time the first leg of the race—the 1.5-kilometer swim—was over, Sarah was right where she expected to be, at the head of the pack. But something happened when she stepped on the sand, a seizing up in her leg. She got on her bike, expecting the pain to go away, but it did not. It was happening; this was happening. She felt panic rise, tried to concentrate on the race, but the pain surged and demanded her full attention, until panic and pain merged as one roaring emotion taking over, its terrible energy picking up with every passing minute. She slowly started falling behind, until finally, at about one minute twenty seconds, she careened off to the side of the course and got off her bike.

A news camera on the scene chose to focus on this moment of drama, and the female commentator for the event is heard sounding maternal. She makes a sound, something between "Ooooh" and "Aaaaw." It is a sad sound, full of concern. Sarah is on the ground, legs akimbo, furiously massaging her lower left thigh with the palm of her hand, grimacing. After a few moments, Sarah looks up at her bike as if it might know something she does not. She stands and tries to get on it, then sits back down again. Gets up again. Starts pedaling with only one leg—maybe she can finish the course, at least. "Well done," says the commentator. A few older people on the sidelines pump their fists at her, cheering her on. As spectators and tourists call out encouragement, she bikes for two more minutes, and the camera pans to the rest of the race, but then back to Sarah—she has gotten off the bike once more. In the shot, she bows down and leans her head against the barrier between the racers and the cheering crowd. Her torso is heaving, the red and white stripes of her athletic suit undulating. A young boy—he is maybe twelve—is on the other side of the barrier, his arms resting on its top. He appears overcome by the sight of this human specimen weeping before him. He bows his own head, buries it in his arms.

"Something in me just broke," Sarah would say later; looking back, she felt there had been so much pressure, to somehow make it up to Ben, to seize this moment for everyone, maybe least of all herself.

A few minutes later, Jeannine and Lauren heard the announcement: Sarah was out of the race.

Lauren and her mother left the stands and found a spot under the bleachers where the two could talk and cry. Lauren was already grieving for how she knew Sarah would take this brutal disappointment. Sarah escaped into a medical van to start to manage her pain alone. At home in New Hampshire, watching from the couch, Adam and his father were silent and ashen. Ingrid, Adam's six-year-old daughter, had tears in her eyes. Adam felt sick. "This is part of the dark side of this stuff," Adam told me. At the time Sarah was one of the top triathletes in the world, but Adam sensed that would be no consolation to her in the weeks and months after the race. "We knew she would still see herself as a failure," he said.

For many years, elite athletes were expected to be as mentally tough as they were physically strong—invulnerable, levelheaded, able to withstand the extraordinary psychological pressures required to outcompete virtually everyone else in their arena. With rare exceptions, such as Michael Phelps, the Olympic swimming champion who talked openly in a 2015 *Sports Illustrated* profile about the therapy he received during a period in rehab, they typically kept any mental health struggles private.

In more recent years, that has changed, most decisively, perhaps, in 2021, when Simone Biles, then a four-time Olympic gold medalist, withdrew from several Olympic events in Tokyo to protect her mental health, the same year that Naomi Osaka, then the number-two-ranked tennis player in the world, withdrew from the French Open. Osaka cited bouts of depression and social anxiety that had weighed on her since 2018, when she defeated Serena Williams in the US Open in a fraught match. Losing in the public eye is excruciating, but for some athletes at that level, even winning is not protective.

As early as 2017, Sarah True started talking openly on podcasts and on social media about a period of severe depression that followed her experience in Rio. She had started struggling with anxiety around the age of thirteen, when the pressure of exams could consume her. But the period after the Olympics was the most challenging Sarah endured. Six months after Rio, she turned to her husband as they were driving down a wooded road. "Ben," she told him, "I want to die." Unlike other depressions that she came out of relatively quickly, this one went on for months, unrelenting, stalking her, weighing her down as if she would never look up and see light again. She contemplated swerving her car into oncoming traffic; she researched online how she could take her own life. She felt she had let Ben down; she had let her family down; she was a failure, and they would all be better off without her.

Ben was a loving husband, but he could not handle this alone. Eventually, he called Adam, who brought up the possibility of her speaking to an appropriate doctor about the possibility of medication. Lauren met Sarah at her parents' home, where Sarah shared how she'd been feeling. Lauren held her and wept with her.

Depression stalks Lauren as well, a low-grade threat she believes she holds at bay with intense exercise. Now forty-six, she runs ten miles at a time, another gift of strength that she regards as a necessary coping mechanism. At times Lauren had succumbed to the depression, most significantly during her first pregnancy. Her second novel, *Arcadia,* she said in a 2012 interview in *The Rumpus,* was written in pursuit of a utopia, an escape from the grimness of her own mental state. "I was pregnant with my first son, and I am a terrible pregnant person, the opposite of radiant—I'm anxious and grow depressed—because bringing a child into the world is such a terribly fraught ethical problem," she wrote. "We don't need the pressure of one more person on the Earth. Why would I bring a baby into a world with a future that's so clearly barreling toward disaster?"

She could not talk Sarah out of her depression, she knew; but she could assure her that she would be there beside her as she waited it out, that she would not let her go through it alone. She helped Sarah find professional help, and when Lauren went back to Florida, the two women still spoke on the phone often, sometimes the silence between them long and quiet, the sound of one person caring for another wordlessly. Lauren's husband had been that rock for her. To him, Lauren's anxiety and Sarah's depression were not products of their ambition, but nor were they wholly unrelated. "To be the kind of person who pushes yourself not just to succeed but to be in the top, I don't know, two percent of these very particular kinds of fields?" he said. "There's a level of extremity there. Is there anyone at that kind of level who isn't anxious? I'd be surprised."

Two years after the Rio Olympics, Lauren's short-story collection *Florida* was published. In one story, an older sister, a fragile child herself, comforts a younger sister when they have been abandoned on an island; in another, a woman feels strongly protective of her niece, whom she loves, as she says, like a younger sister. In *The Vaster Wilds*, the author's empathy for the bravery of a young girl who is running for her life is boundless; she sees her pain and her beauty with a watchful and observant eye.

The love of an older sister, in her telling, is fierce and almost desperately deep.

During Sarah's darkest months, she went to her parents' home as often as she could, visited with her nieces and nephews, smiling, going through the motions of happiness as if, with enough practice, she would get the hang of it once more. Her parents did not know just how bad things were for Sarah, but they helped her. It is how family often works: They are strangers to themselves, unsure how much their shared pasts, as they remember them, even line up with the facts. The best a family can do sometimes is show up and say: We are here, and you belong.

Sarah eventually came out of her depression. For a little while, in the years following Rio, she was thinking about writing a book, she told me, about the experiences of elite endurance athletes, and not just the ones who win—it would be a book about all of it, the emotional extremes, the loneliness, the requisite self-absorption, the thrill of defying expectations, the human cost of striving for excellence, and, yes, sometimes the depression. But it was a daunting prospect, especially given the example of Lauren's exhausting methodology (and, perhaps, her outsize success). "I don't know that I would dare," Sarah said. "It feels like scaling Mount Everest."

On one occasion when we spoke at her home in Hanover, New Hampshire, back in 2017, Sarah was sitting at her kitchen counter as the light took on a late-afternoon richness, and shadows filled in the planes of her face. "Writing's just a lot of work," she said. "It's really hard." And then she winced. She reached her arm down to her calf. "I just cramped up, sorry." She massaged the muscle. "Yeah, weird." Just the thought of competing with her sister seemed to conjure a long-ago but familiar physical pain, seeping through her ligaments.

The Olympic-length triathlon was never the same for her after Rio—it was the sport that broke her heart. But when she was coming out of her depression, she said, she woke up one morning with a revelation: She was ready to move on from the triathlon and to shift her attention to a three-part race considered even more grueling (more Groffy, maybe): the Ironman. "It would have nothing to do with the results, and purely to do with the experience—the process of getting ready for and completing the distance, not even racing," she told me. It would be more physically demanding in many ways—but it held the possibility of something fresh, untarnished.

In 2018, just two years after Rio, she ran her first world championship Ironman, the 140-mile race in Kona, Hawaii, coming in fourth, a triumph for someone so new to that distance. But her athletic career did not surge in a resplendent comeback; instead, it faltered, repeatedly and publicly. She had flashes of success and then devastating collapses: The very next year in Frankfurt, at the Ironman European Championship, she was on course to win with a seven-minute lead; but with the finish line in sight, she started staggering, her body shutting down, then collapsing. At her second Ironman in Kona, sick with a respiratory infection, she started losing consciousness on the bike and pulled over to the side of the road, unable to finish the race.

For many years, Sarah True was an endurance athlete whose

DNFs—her "did not finishes"—were as well-known as her wins. But over the same years she was struggling to figure out her collapses—Sarah said her medical team's "best guess" was an autonomic disorder, a phenomenon neurological in nature—she embodied a different kind of strength, going public about her struggles with depression, four years before Osaka normalized the topic in the international press.

All athletes perform through pain in public, requiring tremendous vulnerability along with near exceptional strength. Sarah extended that public exposure to her own psychological battles as well, and, in doing so—in discussing her depression openly, on podcasts, in dozens of interviews, in frank tweets—she reached a different kind of standing in the field, as one of a team of athletes who were even then serving as trailblazers in destigmatizing a public discussion of mental health disorders. Mitchell Greene, a clinical and sports psychologist, recalls giving interviews as early as 2018 in which he expressed how much he respected the rare elite athletes, including Sarah, who were starting to speak openly about their depression. Later, the official website of the US Olympic teams recognized, on its home page, those athletes talking about their mental health, including Sarah: "They're talking about it unsolicited. They aren't coming out because they had to . . . or because they're retired now and feel safer talking about it. They're willingly talking about this in the prime of their careers because they want to and think it's important."

At a certain point in time, Sarah True's story could have been seen as a cautionary tale about the hazards of pushing oneself too hard. Or maybe it would have been a story about the consolation of sharing one's humanity, about working so hard, taking such risk, and sometimes failing anyway—a conversation that is itself a kind of bravery.

"Failure," Lauren Groff told an audience of rapt students at Harvard in 2019, "is your friend. I know some of you have never heard this before and never failed at anything in your lives, but when you are creating a work of art, or trying to create a work of art, what you want to do is fail, I think, because what happens is you come up against the boundaries of what you can do and cannot do, the boundaries of what you understand and what you don't understand. And understanding that, you are either able to skirt it or to move the borders." Making something new, something possibly great, all the while risking failing, and failing publicly—all of that is ultimately very Groffy. "Moving the walls is hard work," Groff told the audience, "and it's beautiful."

In 2024, Lauren took another risk, opening a bookstore with her husband in Gainesville, Florida, called The Lynx, a risky business proposition under any circumstances, but especially so for a bookstore such as hers, which highlights banned books and focuses on marginalized writers. (Jeannine did not move its walls—but she did paint and wallpaper them with remarkable skill and efficiency.)

Failure is still part of the Groffs' story—all the full drafts of novels that Lauren writes and literally burns or locks away, the businesses that Adam has started and abandoned, the truly public disappointments and physical struggles as well as the persistence and remarkable reinvention for which Sarah is well known in the world of elite sports. The Groffs' lives kept moving forward, the walls shifting. Sarah, taking time away from racing, got pregnant and gave birth to a son, in 2021; just one year later, she won her first Ironman, at Lake Placid, New York. In July 2023, she won at the Ironman Frankfurt European Championships.

Winning didn't resolve all the pain Sarah True had already experienced; it didn't make her family or friends, already loving, love her more or make her son more beloved. When an interviewer asked her how she felt about her win at Frankfurt, a race that had been so painful in a previous year, she gave an answer that was decidedly neutral, as if giving too much power to winning would also give too much power to losing. "It's definitely nicer to finish the race this time," she said.

When it came time for Sarah to enjoy an award ceremony in 2022, after she won the Ironman 70.3 Eagleman, she was already on her way home, where her husband and child were waiting. She also had to get to class, since she had started pursuing a doctorate in clinical psychology. She didn't want to waste the money on another night in a hotel room, when she had a new obsession in mind: saving for her son's college education. She was looking toward his future as much as her own.

The question her parents so often posed to her and her siblings still resounds in their lives, provoking, pushing, encouraging: "What's next?"

TWO

Generators

All three Brontë sisters had strong opinions about childrearing, and their writings are peppered with implicit or explicit thoughts on the subject. None of them ever had children, but all three taught (Emily briefly), and Charlotte and Anne worked as governesses in private homes. Anne clearly couldn't bear parents who overindulged unruly or cruel children, as reflected in *Agnes Grey*. And Emily, in particular, seemed interested in exploring, through her fiction, how much of children's upbringing shaped their character, using the plot device of adoption to raise questions of nature and nurture. "Now, my bonny lad, you are *mine!*" *Wuthering Heights*' Heathcliff says, with vengeance, when he assumes the care of a baby whose father denied Heathcliff an education. "And we'll see if one tree won't grow as crooked as another, with the same wind to twist it!"

On the question of nature versus nurture, Charlotte Brontë seemed inclined to mock oversimplifications. "If you are cast in a different mould to the majority, it's no merit of yours. Nature did it," Jane Eyre's new boss, Rochester, tells her, as if he believed people came fully formed into the world—but the declaration has the unconvincing ring of someone trying to claw back a compliment.

In regards to her talent, however, Charlotte seemed drawn to the notion that her gifts were the product of nature—that her talent was innate, or at least God-given. Writing to a friend that her imagination saved her at times of despair, she added, "I am thankful to God, who gave me the faculty; and it is for me a part of my religion to defend this gift and profit by its possession." At an early age, she declared herself exceptional, as if from birth: "Edited by the genius, C.B." she wrote on the first page of an elaborate magazine that she and her younger brother, Branwell, created.

But by the time Charlotte, at age thirty, actually sat down to write the first novel that she would publish, *Jane Eyre,* she had more than twenty years' worth of impassioned writing and reading behind her, work that was play, but that could be considered decades' worth of practice for what was to come. As children, she and Branwell had covered countless tiny books in even tinier scribblings, creating worlds, drawing maps, dashing off news coverage of their various heroes' battles, writing with a "creative power carried to the verge of insanity," according to Charlotte Brontë's biographer, Elizabeth Gaskell, who perused the children's various sewn-together booklets while conducting research. The writing had all been fantasy and pleasure, but it reflected countless hours of acquired experience and industry.

Francis Galton, a researcher and writer of the late nineteenth century, a former child prodigy and a polymath—he pioneered fingerprinting and conceptualized key statistical concepts—is credited with first using *nature* and *nurture* as terms to capture the forces that shape identity. Galton, too, looked to adoptions as a way of exploring how much children's upbringing could shape the expression of their traits and talents (he was interested in studying the adopted sons of popes, who for a period had a tradition of taking on the care of their nephews). Galton, who also pioneered crude versions of twin studies, was particularly interested in the notion of genius and how it was transmitted through families, ideas that would later curdle into a set of pseudoscientific beliefs about race and heredity that he gave the name *eugenics*.

More than 150 years after he published his 1869 book on the topic, *Hereditary Genius*, researchers, parents, and teachers, are still puzzling over the makings of talent, offering pronouncements about its origins, how to cultivate it and the circumstances that allow it to thrive.

"Talent is one of the biggest lies ever perpetrated," Ellis Marsalis, Jr., told the journalist Leslie Gourse in an interview for *Skain's Domain*, a biography of his son, Wynton Marsalis, widely considered one of the greatest jazz musicians of his time. Ellis Marsalis, a jazz educator and musician recognized in his own right, was often asked about the nature of talent, having also raised Branford, a saxophone player, Delfeayo, a trombonist and producer, and Jason, a drummer, all of whom have been successful in the field. (Another brother, Ellis Marsalis III, is a poet and computer consultant and another brother, Mboya, has autism and is nonverbal.) "Talent is like the battery in a car," Ellis, who died in 2020, told *The Atlantic* in a profile of Wynton. "It'll get you started, but if the generator is bad, you don't go very far."

Ellis's sons were in full possession of good generators, by which Ellis meant the self-discipline to acquire serious skills. "If you want to be different, you got to do something different," Ellis told his kids, according to an interview Wynton gave in *The New York Times Magazine*. But Wynton didn't interpret that to mean that artistic innovation would be what made his career; he explained that to him, being "different" meant having a more powerful work ethic than everyone else. "If you practice an hour a day, you'll be like everybody who practices an hour a day," he said. "If you want to be great, you be the one doing five hours a day." The interviewer asked him if his father put a lot of pressure on him to succeed. "No," Wynton said when asked. "My father never put pressure on me. He's too cool for that kind of stuff."

A childhood friend, quoted in *Skain's Domain*, remembered that discipline kicking in for Wynton in middle school. "He would wake up in the mornings and practice before he went to school and many times afterward," recalled Victor Goines, a jazz saxophonist and clarinetist who grew up, like the Marsalis siblings, in New Orleans. "I went by the house, and he was practicing." Wynton told the American Academy of Achievement in 1991 that when he was coming up as a musician, "I went about seven years without missing a day of practice." He had a strict regimen: "I would not go to sleep until I had practiced all the stuff I had to practice."

DRIVE AND DISCIPLINE, Ellis Marsalis seemed to suggest, were what made the difference between a good musician and a great one. Given his practice regimen, Wynton Marsalis had clearly internalized his father's values or come to them on his own by middle school. But when *Harvard Business Review* asked him, in 2011, "What's your theory on talent versus practice?" he emphasized something different. "You can become proficient at anything," he said. "If you're a boxer, you can practice four million hours and become proficient to a certain point, but if you don't have the talent, you won't be the one to beat. You can't practice the ability to make connections or have a deep, spiritual insight."

In his comments, Marsalis seemed to be indirectly referencing the idea, popularized by Malcolm Gladwell in his 2008 book, *Outliers,* and oversimplified by many exposed to it, that reaching ten thousand hours of practice was what separated the best from the merely great. In his book, Gladwell tried to refine the premise that achievement was simply "talent plus preparation," citing work done in the early 1990s by K. Anders Ericsson, who later became a psychology professor at Florida State University. "Their research suggests that once a musician has enough ability to get into a top music school, the thing that distinguishes one performer from another is how hard he or she works. That's it," wrote Gladwell. "The people at the top didn't work just a little bit harder," he said; they worked "much, *much* harder." He argued that "the closer psychologists look at the careers of the gifted, the smaller the role innate talent seems to play and the bigger the role preparation seems to play."

At the very top of many fields, success clearly requires both exceptional talent and unusual effort, even if that challenges the much more comforting notion that many seemed intent on taking away from the book—that joining the ranks of the very best was simply a matter of putting in those ten thousand hours of practice. As for the proportions of talent versus effort that are required to reach the highest echelons of a given field, one might sensibly assume that, could they actually be irrefutably parsed, they would likely vary depending on the domain—maybe some specific innate talent matters more for violinists at the top of their field than it does for, say, the most illustrious photographers.

Nor is it a simple matter to disentangle talent from effort, points out David Epstein, the author of the best-selling books *Range* and *The Sports Gene*. Epstein argues that few strivers, without some unusual innate talent, would likely ever put in those ten thousand hours, because of how talent and drive are interconnected. Without some innate, perhaps genetically bestowed advantage—the exceptionally acute eyesight common among baseball players, for example—a young person may not get the feedback that would justify investing that time. The Marsalis brothers likely had some innate musicality inherited from their talented father and their mother (whose extended family was rich in professional jazz musicians). Without that gift, whatever drive they possessed might have found some other outlet or might not ever have even surfaced in their lives. Wynton could not achieve what he has without practice, but without the level of talent he possessed, he might not have been driven to work as hard as he has. "Talented people almost always know full well the excellence that is in them," Charlotte Brontë wrote to a friend in 1846. (Whether they have the opportunity to express it, however, is often a matter of circumstance.)

Social psychologists and personality researchers frequently debate among themselves the question of which personality traits and talents are most linked to high achievement—musical, academic, or otherwise: Is it conscientiousness? Cognitive skills? Is it some combination of conscientiousness and what personality researchers call "openness to experience"—a quality of curiosity that can be channeled toward the arts or abstract ideas? Or is it grit, the formulation that the University of Pennsylvania psychologist Angela Duckworth made famous, a combination, as she defines it, of passion and perseverance over time to achieve long-term goals?

Even if researchers could answer those questions definitively, the utility of the answers would only go so far, because the question of how to shape personality traits remains surprisingly elusive. It is not that personality traits are believed to be immutable, impervious to random environmental forces or concerted efforts intended to mold them in one direction or another; it's just not clear what interventions would effectively shape them with lasting results, with few exceptions.

Many scientists who study measures of intelligence, such as IQ (itself a highly contested concept, critiqued as socially biased) do think that it is, to use a scientific term, "mutable"—it can clearly be changed by some known environmental forces. A 2018 meta-analysis, published in the prestigious journal *Psychological Science,* teased out data to find, across a sample of 600,000, that for every year of education, a person's IQ rises, on average, one to five points, leaving an effect that lasts across one's life span. Social psychologists have struggled, by contrast, to demonstrate how to mold, with lasting effects, what are sometimes called "noncognitive" skills, such as perseverance or creativity.

After Angela Duckworth's concept of grit was the subject of a successful 2013 TED Talk (and three years later, a best-selling book), some schools started incorporating the instruction of grit into their curricula. Districts in California began testing students for character traits such as self-control and conscientiousness, measures that would be factored into the evaluations of school performance. Assuming Duckworth was correct, and grit had such desirable properties, shouldn't the educational system find a systematic way to cultivate it?

Long before California districts launched that initiative, Angela Duckworth worked with the KIPP charter school in New York City, which educates mostly disadvantaged youth from low-income backgrounds, to help create what they called a Character Growth Card—a way of reflecting back to students, through questionnaires, feedback focusing on certain behaviors, such as those that required self-control. The school incorporated messages about character into lesson plans and encouraged teachers and parents to model strong character traits, such as perseverance. In studies of middle schoolers who attended KIPP, the school's students surpassed others in their demographic on academic achievement and persistence in continuing their education.

But for all that focus on character, noncognitive skills did not seem to show much change. Compared with peers who had entered the lottery, but hadn't been randomly selected to enter various KIPP schools, students who did attend were no more motivated or higher in self-control, an independent 2013 study found; if anything, the KIPP students were more likely to self-report themselves engaging in "undesirable behaviors," such as being rebellious in class or defying a teacher. (The researchers did point out that precisely because of the high standards the KIPP students were being taught to uphold, they might have been unusually hard on themselves in self-evaluations.)

In a 2016 editorial in *The New York Times* titled, "Don't Grade Schools on Grit," Duckworth pointed to research showing the benefits of teaching children how to set goals and make plans or to adopt a "growth mindset" about intelligence. But still, she did not think the research was far enough along ("and perhaps never will be"), to justify judging teachers and schools on how well they "taught" students character. (She was also wary of holding "accountability" over teachers' heads, as opposed to using information for reflection and feedback.)

As a school system, KIPP, which has since moved on from earlier approaches to character-building, offers an important model for improving students' academic scores and their likeliness to enter and remain in college—which is obviously distinct from changing a young person's character. Whether or not that's even an appropriate goal to pursue is up for debate, suggested Marcus Crede, a psychology professor at Iowa State University raised in South Africa, who has also published research that critiques Duckworth's methodology. "I would be somewhat unhappy as a parent if my daughter came home from school and said her teachers were trying to change her personality," said Crede in an interview published on the blog *Education Next*.

And yet many parents do hope to see in their children some internal motivation—some willingness and ability to pursue a chosen goal with focus and determination. Julia Leonard, a researcher who collaborated with Duckworth during her postdoctoral years, has moved on from the University of Pennsylvania to a lab of her own, the Leonard Learning Lab at Yale, where she has taken up the question of what motivates children to decide it's worth it to keep trying.

By the time Leonard started running that lab as a thirty-one-year-old developmental psychologist, she had been trying for years—twenty years, approximately—to figure out how to make a person more "effortful," or, as she prefers to put it now, how to help children persist through challenges. Her career unofficially started when she was a young girl growing up in Berkeley, the daughter of a mother with a PhD in epidemiology and a father who was a professor. Leonard was always an energetic doer, a child who asked, relentlessly, "Can I?" Could she get out the blocks? Could she take skating lessons? Could she manage to make breakfast herself? Her older sister, Sarah Rose—a girl who was dreamy, who liked to sing or daydream for hours, just thinking her thoughts, imagining stories—was forever frustrating Julia. Sarah Rose would not try to catch a ball, would not join Julia in her elaborate construction projects with blocks or join her in executing new twirls on the ice-skating rink; she tended toward "I can't." This scenario played out over and over again, inspiring worry and sometimes even a little indignation in Julia. How could her sister just give up like that? How could her parents let her? When the girls got older, their family sometimes went skiing. "Look, Sarah, we're racing!" Julia would call out to her sister, who had thought they were really just skiing downhill alongside each other. "And I'm winning!" Julia would shout. Sarah would nod encouragingly—she was an older sister, after all, and nice that way. "But I'd be thinking, 'Well, it is true you are ahead of me, but since we are not racing, you are not winning,'" recalled Sarah Rose. "'And even if you are—that's fine.'"

Julia always tried to be the best, especially at her favorite sport, ice-skating, even when it became clear, as she turned twelve, that she was too tall for the sport, too skittish about the more advanced jumps, and suddenly, also, too anxious before competitions, her heart racing for days before the actual event.

"I just want to come in first once," she told her mother. "And then I'll quit." She worked on a double jump, begged her mother to buy her the kind of glitzy costume the judges liked, and kept it together for one last competition. She came in first. She quit a week later.

Sarah Rose saw a somewhat obscure off-Broadway sequel to *Annie* when she was ten, and for her that was it—the only thing she ever cared about, ever wanted to succeed in was the world of theater. When she was young, at least, her parents never discouraged her. "I think you could work as a janitor in a theater, and you'd be happy," her father once observed, which told her he understood her completely.

And yet even as Sarah Rose valued that one singular passion, she did not always find it easy to see her sister excel at so much. Julia had social ease. She could paint; she was athletic; she got perfect grades; and she was good at card games, board games, and even palm reading, which she learned from a kids' book and kept up for people's entertainment at parties.

Sarah Rose could also perceive what all that energy and exertion took from her sister. "She is such a fun person to be around," she told me. "And then she can flip and get super highly anxious about school or work." Sarah Rose has watched her sister in action for years and wondered about the trade-offs of that kind of drive. "On the one hand, I'm so relieved I don't have that high anxiety and that I don't care that much," she said. "On the other hand, that's what's made her achieve so much. There's a constant question for me, and a lot of other people: Do you need to be riddled with anxiety to make good work? Do you need that pressure?"

Julia's drive continued through college, and then through graduate school in brain and cognitive sciences at MIT, where she still excelled—but also struggled to manage the stress of the work, the uncertainty, at times succumbing to the same torturous, heart-pounding anxiety she felt as a girl in the days before an ice-skating competition. She, like Lauren and Sarah Groff, both benefited and suffered from having a powerful generator, as Ellis Marsalis called it. Its energy was sometimes overpowering, driving her forward but exhausting her system. Her successes allowed her to keep doing work that she loved, but the emotional stakes of the work, her struggle to manage her feelings, made her wonder at times if she could keep going.

If Julia failed, she would fail a small circle of people that included a handful of advisers, and, most important, herself. There would be no harsh critic publishing a withering review in a prominent paper, no camera confronting her to ask her how she felt about a race that she was unable to finish. But she had faced a different risk, which was the possibility that months of her time and tremendous hard work would add up to nothing. Doing research well means deliberately designing a study that fully allows for the possibility of failure. A hypothesis disproven is no failure for science because it provides more data, a theory eliminated; but the absence of results often feels like failure for the scientist. So much scientific research entails spending weeks, or even years, waiting for results, running on fumes of hope that are mixed with dread.

In graduate school, to relieve the stress of pursuing a PhD, Julia took up rock climbing. As she watched climbers engage in friendly competition, she found herself reflecting once again on the question that had been with her since childhood: What makes someone try hard? When she saw someone at her level struggle but eventually succeed on a climb, she noticed that it pushed her somehow—it made her feel better prepared to try harder as well. Julia, who was working at the time with fifteen-month-old babies, started to wonder what, if any, cues babies that age would take when they saw an adult try hard and succeed at some small task, as opposed to seeing that adult succeed with ease. Would the adult who exerted effort inspire those babies to try harder with a challenge of their own?

Julia, at home in her bedroom in Cambridge, rigged a toy with somewhat mysterious properties, designed to test the persistence of fifteen-month-old babies. She purchased a soft cube with a large, brightly decorated button on its surface, which she disabled. Then she she affixed onto another surface of the cube a sound module (taken from a greeting card) that played "Twinkle, Twinkle, Little Star." Then she made it hard for the babies to find the module: She covered the cube, including the sound module, with a layer of soft felt. The gadget that would play the song would not be obvious to the babies, although the adult experimenters who demonstrated what it could do would be clued in. She left exposed, as an enticement to effort, the large, brightly colored button that she had disabled. Her ultimate goal: to see whether babies could be inspired by something in their environment to try harder to get results from that toy—to persist in trying to get a result (the reward of the music) from a toy designed to confound them.

That summer of 2014, Julia started spending many of her weekends cruising a toddler play area at the Boston Children's Museum, looking for babies around a year old whose parents were willing to cooperate. Julia hoped to prove that the babies who watched an adult succeed after appearing to struggle with a small task, such as getting a toy out of a container, would themselves persist longer in trying to make that large button on their own toy play "Twinkle, Twinkle, Little Star." Perhaps a certain kind of modeling of effort would elicit a different response: As rational, learning machines, they would see the cause (effort) and note the effect (reward). They might internalize the connection, extrapolating that reasoning to a different toy from the one that the adult researcher had managed to manipulate.

Working with babies is notoriously challenging, but Julia was too persistent to let that give her pause. She ended up conducting three studies on this topic on around 260 babies, an unusually high number for a trial like this, given how difficult it is to recruit participants that young. The results were intriguing: The babies who watched Julia or another experimenter struggle and then succeed were, in fact, the ones who pushed the button significantly more times in their (fruitless) efforts to elicit the sound of music. The paper, meticulously researched over several agonizing months, was an early professional success for Julia, published in the prestigious journal *Science* in 2017. "The fact that [babies] can get these messages just from watching someone and then internalizing it tells us that if we want to be raising children who are persistent—who have what you might call 'grit'—we ought to be starting early and we ought to be thinking about how we model our own approach to challenges," Lucas Butler, a developmental psychologist at the University of Maryland, College Park, told *Scientific American*.

In 2019, Julia followed up her research with work with four- and five-year-olds, pairing the modeling actions the adults did in front of the children with language—either a pep talk ("You can do this!") or expectation setting ("This will be hard") or value stating ("Trying hard is important"). In a control group, a parent said nothing. In the end, Julia found that children's persistence was highest when the "the adult both succeeded and practiced what she preached: exerting effort while testifying to its value." It wasn't enough just to say, "Trying hard is important"; the children were most inclined to try hard themselves when they saw the beneficial results of a parent who put that principle to work.

When Julia told her family what she was working on, at the outset of her research, her sister Sarah Rose confronted her. "This work is totally about me, isn't it?" she asked her. At the time, Sarah Rose had graduated from NYU but was going from part-time theater job to part-time theater job, babysitting to cover her rent. Julia got annoyed, and the two squabbled back and forth about it a bit. This had nothing to do with her sister, Julia insisted. Why would she assume that? Was everything about her? If anything, this inquiry, Julia argued, was inspired by her own observations about herself—about what motivated her to push herself when she was rock climbing. And, more important, it was about trying to find interventions that could improve long-term outcomes, especially for disadvantaged youth.

But when the study in *Science* was published, Julia could finally admit to her sister, and even laugh about it: Yes, it was true—although the work's importance was in its potential to help children have better outcomes, probably the research direction was somewhat inspired by the frustrations she had felt about her sister when they were young.

Part of what made the conversation easy to have was that it was not just Julia whose long-term effort had paid off with gratifying results. All the time that Sarah Rose had been daydreaming or blowing off math to read plays in her bedroom at home, all the time that she had spent working as a nanny during the mornings and working for pennies at night in small theaters, had turned out to be its own kind of grit, a kind that Julia might not have recognized because it looked heavy on passion and light on consistency. In her early twenties, Sarah Rose had been strategically taking on small directing, producing, and dramaturgy projects wherever she could. When she was twenty-seven, she landed her dream job, one that would allow her to call upon the expertise she had acquired over so many years: She became the literary manager for Berkeley Repertory Theatre, a highly regarded theater in one of the country's most culturally rich small cities. Doing work she found stimulating and creative, she was, by any measure in her field, thriving; now Sarah Rose works as a freelance dramaturge at various theaters while running live events for the Bay Area's public radio station, KQED.

In part because of how she had seen her sister's life unfold, Julia had questions about her research from time to time—questions that went beyond whether its results would translate beyond early childhood, if a parent repeatedly consistently modeled a million small moments of effort. "This is something I come up against a lot in my own research, where people are like: You know, you study persistence. But is it good to persist? *Should* you persist? Isn't it also good to give up? And what is it that you really want for kids? Do you want them to just work, work, work, work? Or is there any merit to pleasure?" As she thought about her work over time, she started thinking less about the results—the effort someone's willing to put in—and more about the starting point: What makes a person want to try—to have that exciting energy of inspired effort? How can adults foster in kids a sense of agency, the feeling that they can accomplish their own goals? And what effect might parents have on that intrinsic motivation?

Various studies, over the years, have shown that when parents intervene, for example, as a child is trying to finish a puzzle, the extra help is demotivating for that child. But it was always hard for researchers to separate cause and effect—maybe these overly involved parents intervened because they already knew their children were not going to try that hard on their own. Julia designed a study in which experimenters, having been randomly assigned to various children, either intervened in a child's given task or did not—and she found that the presumed effect held. "Having an adult intervene was extremely demotivating," she said. "It had much more of a negative effect than any of the positive effects my other studies had"—the ones that encouraged persistence. In another iteration of the experiment, kids were told the experimenters were intervening because it was a turn-taking game. Even when the kids didn't think the adult's intervention reflected a lack of faith in their ability, the kids tried less hard on the next task or game they were asked to undertake. It was also hard to reverse that demotivation in the near term—for example, with subsequent games within hours after the first.

Julia wondered why parents were intervening in the first place. Maybe some parents were simply more or less inclined to believe in their children's abilities, independent of what those skills were. Could she change their beliefs about the value of letting kids try to execute daily life challenges on their own, and, if so, would that change parents' behavior? She and colleagues started by conducting research asking parents to identify the everyday situations in which they found themselves taking over a task for their young children. The answer turned out to be, for parents of four- or five-year-olds, simple routine tasks such as getting dressed.

"It turned out that parents didn't think kids learned much from mundane daily activities like putting on clothes or cleaning up, compared to more academic tasks like completing a puzzle," Julia said. She found this concerning. "My prior work shows that when you take over for one task, it demotivates kids on the next task, even if it's totally different—it could have a spillover effect," she said. "And you build habits not just for kids but parents—it's not ideal for them to be accustomed to taking over for their kids without realizing it might be taking away their autonomy."

Julia was inspired to design another study, one that took place at the Please Touch Museum in Philadelphia. Parents of four- and five-year-olds were offered the opportunity to have their children engage in some play that required those children to wear hockey gear. The point of the exercise, for Julia, was never about hockey, but to see if the experimenters could influence whether or not parents intervened as their children put on unfamiliar gear ("Thank God we don't work in Canada," she said). All the children had to put on a chest piece that needed to go on over their heads; they needed to use their hands to strap the sides. They also had to wear large shin guards that required them to use one hand to hold the piece of equipment to their shin, and the other to fasten it. "They can do it," said Julia, who knew that was the case because she had already conducted a study in which her team directed parents not to help at all—and, to the parents' shock, the children succeeded. Now the question was: Could parents be inspired to let the children try to get themselves dressed without being told explicitly to stand back and let them figure it out?

The answer, she found, depended heavily on what messages parents received in the moment. Some parents, who received a one-page piece of paper about the day's activity at the museum, were left with a parting thought at the end of the write-up that was direct: "Fun fact: Children learn a lot from putting on clothes themselves! They learn things like motor skills, problem-solving, and self-confidence. The skills children learn from putting on clothes help them develop into capable, independent adults." Other parents were left with a neutral parting thought that was considered the "control" in the trial: "Fun fact: Many museums have dress-up stations for children, where they get to put on various clothes and outfits themselves. Putting on these clothes helps children interact with the museum and engage with the exhibit. In each of the rooms was a mirror with a photo of how the uniform should look so the child could see what the end goal was." Which message the parents received apparently made a difference: The parents who were told that their children might learn from dressing themselves let them do so at a significantly higher rate than those who received the neutral message about museums and dress-up.

For parents invested in their children's success, it can be excruciating to watch them fail—or even to wait it out and see whether their children will succeed on their own. But Julia's research suggests that parents' desire to help children learn is ultimately more salient than their expectation or fear that they might fail—they just need to be encouraged to give their kids a chance to succeed or fail on their own. Parents need reminding that the best way to let kids learn—to help them succeed, ultimately—is largely to allow them to risk floundering, especially if they're eager to take the challenge on. Learning to struggle is an education in and of itself—but also, the child may well exceed the parent's expectations. A young person might be more capable of getting out the door for school without a personal wake-up than a parent thought; she might break the women's record for fastest swim of the local lake.

Julia is now in her mid-thirties and thinks about becoming a parent herself in the near future. She said she likes to think that she'd be the kind of parent who lets her child risk pain or disappointment—the kind of parent who messages that approach early, who lets a child fumble around with the Velcro or the shoelaces or the last few puzzle pieces before offering a hand. But she is also an ambitious person, likely to be a working mother—and she recognizes that it takes work and patience to let a child struggle. It takes time, and sometimes support, and enough resources that parents aren't overwhelmed simply trying to meet their children's basic needs. She has great sympathy for the parents who, in the rush to get out the door, simply do not have time to allow a child to puzzle over the backpack closing clip as the minutes tick by.

I felt she understood and had sympathy for a parent, for example, like me—someone who has probably spent weeks of my life, all told, searching for stray toys, then misplaced school folders, then missing AirPods, rather than waiting out my children's distress as they searched, less efficiently, for those missing objects themselves. Things needed to get done; sometimes I just did them myself.

"The skills children learn from [fill in the blank] on their own help them develop into capable, independent adults"—we all, in my family, could have benefited from a few posters around the house bearing that message. Julia does believe that small interventions could remind parents to let their children try, fail, try again, and ultimately learn. "Imagine if there were signs all around grocery stores," she said. The signs might read: "Did you know shopping is learning?" Why not encourage parents to give their children tasks—to come back with eight apples and nine pears or to go in search of something called a pinto bean? "It's so simple," Julia said.

As we spoke, I flashed back to a small bit of ingenuity I'd admired when I visited the Groffs in New Hampshire: Jeannine had put glasses in the bottom cabinets, which usually held pots, so that her grandchildren could be self-sufficient. I thought about her tiny granddaughter, in a cast, hauling herself up the stairs. I remembered a quiet moment at Adam and Tricia's home when the dog started whining and Tricia Groff took a quick break from our conversation to turn to her daughter, then six. "Ingrid, can you put the dog on the outside chain?" she asked, her voice gentle but intent. "Can you do it? Can you do it? Go ahead and try." It was a brief, passing exchange—but I also noted that she resumed our conversation without pausing to see how her daughter fared with this clearly new task. She trusted her daughter to execute. I imagine that an accretion of those small daily interactions could build a deep reservoir of confidence and competence.

Asking a parent to know how far to trust their children's capability seems simple—and, at the same time, Julia wonders what's even reasonable to impose on parents. "Would I think about learning, as a parent?" she said. "I'd probably just think about keeping my kid safe. And getting through the day. Sometimes I wonder if I'm asking too much of parents."

Ultimately, all of Julia's striving and worrying and working and experimenting was rewarded with a meaningful reward: She was offered the opportunity to run her own lab, the one she now heads at Yale. And yet she did not experience that transition, which happened in 2020, as entirely one of triumph, because of how the pandemic affected her. In some ways, by the end of those long years, she told me, she felt less confident than ever—more humbled by her own limitations. While some people used the pandemic to learn to bake bread and start running six miles every day, Julia, usually such a doer—a rock climber, an accomplished potter, in her soul something of an artist—lost all of her motivation. She had all the time in the world to think creative thoughts, to make art, to come up with new scientific ideas, but none of that flowed. "I wasn't the person I thought I was," she said. She was forced to confront how much of the drive she considered innate was dependent on specific circumstances, predicated on certain conditions being available to her—a lab to visit every day, people with whom to discuss her ideas. She was always aware of the privileges that allowed her to succeed; but she still emerged from the pandemic less convinced that she was some kind of irrepressible overcomer, the kind of person who would persevere and succeed regardless of the hurdles put before her.

Following the pandemic, her orientation—and that of much of her field—has shifted away from notions of grit and clear-cut success and toward a less goal-oriented stance. A generation of young people seem less interested in cultivating their own grit than in questioning the underpinnings of a society that for so long valued, above all, the kind of work effort that requires long hours, often to the detriment of overall well-being.

"The world has changed," Julia said. "Science has changed. I've changed." Rather than emphasizing academic achievement, she and many in her field are finding new energy in pursuing efforts to preserve the natural love of learning and curiosity that children seem to lose over time; instead of seeing a preoccupation with grit, she sees a research shift toward adaptability. "I also see a shift to circumstance," said Julia. "We have a range in who we can be, and the key is figuring out what circumstances will push individuals to the upper limits of their own range."

Julia and her sister still live on opposite coasts, each of them pursuing their professional passions, sometimes to the benefit of their personal lives, sometimes to the detriment. When the two sisters are together at a restaurant, Sarah Rose, who always struggled with math, still frustrates Julia by handing her the check when it's time to calculate the tip. Julia still wonders: Can't she even try? Even so, Julia said she's become much more curious about her sister's natural strengths, less interested in figuring out how best to "fix" her.

And Sarah Rose still knows that, for her sister, there is a flip side to that unusual kind of drive, which is the anxiousness that comes with it—the restlessness, the worry, the stress. Julia's research suggested that, at least in babies, the modeling of gritty behavior can be motivating. It's tough being tough; it's up to our children to decide when it's worth it.

THREE

The Holifields

As a girl growing up in Tallahassee, Florida, in the late 1950s and early 1960s, Marilyn Holifield had two great passions. She loved playing the piano, and she loved swimming, and she couldn't have said which one she liked better, although swimming did have one distinct advantage. Over the summer, she spent so much time in the water—hours of laps in the morning and hours in the afternoon—that her mother gave up on taking her to have her hair straightened, temporarily liberating Marilyn from the sizzle and smoke of the hot comb and curling iron. Those long summer days enjoying her natural hair contributed to an appreciation of her town pool, named Robinson-Trueblood, as a place where she felt free, even if, in fact, the destination was imposed on her—the other two pools in town were strictly for white people.

For Black families, Robinson-Trueblood Pool was one of Tallahassee's social centers, a place where teenage girls competed in beauty contests, where local churches held baptisms, where older kids held dance parties. Sometimes the entry line grew long, and the parking lot became a hangout scene in its own right, accompanied by the tinny sounds of car radios playing Smokey Robinson or the Supremes. Marilyn enjoyed all of that, but she was primarily at the pool to swim, one of only three girls on the swim team. She even swam faster than her older brother Eddie, who was one year her senior, at least until he refused, piqued, to keep racing her. Their brother Bishop was a year older than Eddie, and she could beat him, too, but he seemed to accept this with fraternal pride. Even now, Bishop marvels that "her arms just moved faster than everyone else's."

Two days after the Civil Rights Act passed in July 1964, when Marilyn was sixteen, the city closed Robinson-Trueblood, along with the city's other two public pools. The mayor claimed that the pools were being shut for financial reasons. But everyone knew the real story: Earlier in the day, several Black people had tried to integrate one of the two white pools in town. Tallahassee was one of many cities and towns, especially in the South, where officials opted to drain and close public pools in those years, rather than follow the new laws of the land.

The closure was demoralizing for the Black community in Tallahassee, one more cruel racist offense, but for the Holifields, the loss hit especially hard, because their connection to the pool—to swimming—was so deeply intertwined with a family tragedy.

In May 1946, two years before Marilyn was born, some family friends asked her parents if their three-year-old daughter, Gail, wanted to join them on a picnic. The couple who offered the invitation had a fishing boat, and the husband was leading a group to a small lake nearby. At the time, many Black families owned land by that lake, and others went there to swim on hot summer days or for celebrations, roasting hot dogs by the shore.

Not long after lunchtime, the local sheriff received a call: an emergency at the lake. One child had stood up, another reached over to seat the child back down, and then the balance suddenly shifted; the boat capsized, spilling everyone into the water, just twenty-five yards from shore. The teacher who had taken them out on the boat managed to make it to dry land, along with his own seven-year-old daughter, who was clinging to his neck. But he had lost his son. As his wife wrung her hands and paced along the shore, divers arrived to search the water for the six other children, none of whom survived. One of those children was Gail Holifield.

In the Holifield family, the aftermath of the tragedy was like the lake itself: quiet, deep, and large. Some saw a permanent change in the siblings' mother, Millicent Holifield. "She never got over the loss of that little girl," said Carriemae Marquess, who showed up at the lake that day and was a friend of Millicent and her husband, Bishop Sr. "I think she carried a sadness with her. She never spoke about it, and nobody ever asked her about it."

The death of any child is a particular kind of horror. In this case, though, the Holifields' loss was bound up with another horror, the fraught history of how racism denied Black people the chance to swim in safety. As Kevin Dawson details in his book *Undercurrents of Power*, Black children, barred access to pools and beaches staffed by trained lifeguards in the Jim Crow South, often swam in unguarded rivers, lakes, and swimming holes, which meant that they drowned at higher rates than their white counterparts. Even now, Black children drown at nearly three times the rate of white children.

After a tragedy like the one the Holifield parents endured, many parents might have warned their children away from deep water. But that was not how the couple responded. Millicent decided instead that her children would swim as well as anyone, Black or white—that they would never be in danger of drowning. It was expected that they would all take swimming lessons, that they would join the swim team. They would all three learn how to be lifeguards. "Fear," Marilyn said, "was not their thing."

By 1964, when the city of Tallahassee closed its public pools, the Holifields had had many conversations about the student protest movement gaining momentum throughout the South; over dinner, they discussed the marches they watched on television, the brave nonviolent direct actions, the sometimes deadly retaliations, the rising national support for protesters, the harbingers of long-awaited progress. Marilyn, Ed, and Bishop had supported the civil rights movement, but they hadn't become activists as a family—until Robinson-Trueblood closed.

On June 23, 1966, the local paper reported that three Black protesters and one white one had showed up to picket City Hall, in response to calls for activism from a group that the article hinted had some outside influence. Who that white woman was is lost to history, but the remaining three were Ed, Marilyn, and Bishop Holifield, representing no organization other than their own family. Their father considered himself "the head of the transportation committee," picking them up in the afternoon, but the project was Bishop's initiative. "My bathtub's not big enough," read Marilyn's sign.

It took another year before the question of whether to open the pool was finally put to a straw poll. By then, Bishop had just finished his first year of law school and had secured a summer internship with the Law Students Civil Rights Research Council chapter back in Tallahassee. For his internship, Bishop launched a stealth phone campaign, relying on Ed and other volunteers, to make sure that those in favor of the pool's opening—mostly Democrats and Black voters— showed up to vote. On August 1, 1967, three years after the pool was closed, Bishop stood in the gleaming lobby of the newly built City Hall, one of a handful of spectators overseeing the tallying of the votes, watching the two piles of ballots climb at nearly the same pace—until it was finally announced that the vote to open the pools had won by a narrow margin of about five hundred votes. An undertaking to reopen the pools would be placed into the city budget. It was a moral victory, but for the Holifields, it was also a deeply personal one. A baby drowns; some twenty years later, siblings who never knew her became civil rights activists, the fight to reopen swimming pools the first of many forms of resistance their activism would take over the years to come.

In their professional lives, the siblings shared much common ground, with each committed to civil rights; but how they each leveraged their respective power was also distinct. Bishop founded an influential Black organization at Harvard Law School, then devoted years of his life, as general counsel of a historically Black college in Tallahassee, to reinstating its law school. Marilyn spent two years of high school and periods of her professional life somewhat isolated, rarely deterred, helping to desegregate more than one powerful institution, eventually becoming a civic leader and the first Black female partner at a major firm in Florida. To this day, she is one of a vanishingly small number of Black female law partners—they represent less than 1 percent of the total—at major law firms around the country. Ed, a cardiologist with a master's degree in economics, eventually left the workforce, channeling his outrage, with mixed results, as a public health community activist, an unrelenting combatant, a bristling figure who took pride, at times, in being a thorn in the side of various institutions. Their parents, who led lives in which success was itself a form of protest, primed them for the years the siblings spent taking on one challenge or another; but as I so often found in my reporting on siblings, they were also motivated by one another's actions, keeping an eye on whatever bar another sibling was setting on his or her own.

Their story stands out not just for their sheer talent or their drive, but for the ways they, like their parents, intersected with the currents of history, lifted by the powerful institutions they'd fought to access, and opening up opportunities for others along the way.

THE FORWARD PROPULSION of the Holifield family may have started with a fall: Their father, Bishop Holifield, Sr., born in 1909, tumbled from a tree as a child. His family, living in Hillsboro, Mississippi, an agricultural corner of Scott County, was unable to find the medical help he so badly needed, and he walked with a pronounced limp for the rest of his life. Maybe because of that injury, a cousin of the siblings theorized to me, Bishop Sr. set his sights on a college degree, the only one of the family's thirteen siblings to do so, compelled by the concern that he would never excel at physical labor. Marilyn believes the motivation was more intrinsic—that her father was someone who wanted a bigger life than what he saw all around him, even as an adolescent.

A high school education was not available to Bishop Sr. in Hillsboro, which is why he left the five-hundred-acre farm his family owned and set his sights on the Tuskegee Normal and Industrial Institute. That school would evolve into what's known as the Tuskegee Institute, an HBCU closely associated with its founding principal, Booker T. Washington. To get there, he'd have to travel some 250 miles east by "Jim Crow coach," a dirty compartment with no amenities where white travelers sometimes transported their chickens and hogs.

Bishop Sr. was an overcomer, a man who had the bearing of a man taller than he was. At Tuskegee, he thrived, studying under George Washington Carver, the most celebrated Black scientist of his time, working with him on his famed research on the many uses of the peanut. He was especially pleased to be working on the extraction of peanut oil because, as he told the *Tallahassee Democrat* years later, he was losing his hair at the time and peanut oil was thought to counteract hair loss. (Alas, he added: "It didn't.")

Bishop Sr. was, in fact, older than might be expected for a college student: To pay his tuition, he alternated some years of schooling with years of work, selling fruits and vegetables he grew himself. He finally graduated in his late twenties from the Tuskegee Institute with a bachelor of science degree in agronomy and soil science, and, by 1939 landed a job as the chief landscaper at a nearby veterans hospital. He went on to become a water and soil conservationist, building a career based on the premise that, with the right tools and skills, growth could happen under even the most hostile conditions. It required persistence; it required knowledge; it required, especially, long-term vision, the patience for incremental progress—the expectation that, over time, one's efforts would eventually be rewarded.

While working at the veterans hospital, he met Millicent Clark, then a nurse working at the same hospital. Millicent was raised in Boston, the daughter of a working-class father from Barbados and an activist mother from Suriname, parents who raised her on the principles of Marcus Garvey, a Jamaican proponent of self-reliance who championed a powerful sense of Black pride in the early twentieth century. Bishop Sr.'s Tuskegee education, shaped by the philosophy of Booker T. Washington, instilled in him the imperative for self-advancement. Their respective ideological educations were different in some key regards, but had enough in common that her parents were extremely compatible, Marilyn said: "They came from two strong influences that promoted financial independence." Bishop Sr. was an ambitious child of the South; Millicent, a trailblazer from the North who'd attended Boston City Hospital's nursing school, was one of only three Black women in a sea of white faces captured in a class photo that left a strong impression on her daughter.

Some of Bishop Sr.'s first steps toward financial independence started with a job with the federal government. He was the first soil conservationist in the state of Florida to be hired by the US Department of Agriculture, a role that had him working alongside white employees and educating farmers—Black and white—about best farming practices. The job had the advantage of security and prestige, but his first assignment suggested how profoundly challenging the work would be: He and his new wife were supposed to move to Marianna, Florida, in 1943, a town where, that same year, a Black man had been lynched; Marianna, by then, was already infamous for being the site of the last public-spectacle lynching in 1934.

Fate intervened, in the form of a letter from a local government official. The letter said the authorities in Marianna would not welcome Holifield in that role because his success would "set a bad example" for other Black people in the area, as Marilyn recalled, having read that letter many times as a girl. To Marilyn, it held great power: a thin sheet of paper, limned with hate, that led their father to a new assignment, and led their family on a different path, likely a much brighter one. That piece of paper was the twist of fate that landed them in Tallahassee.

WHEN THE HOLIFIELDS ARRIVED in Tallahassee in 1943, it was a semi-rural, somewhat sleepy state capital, a town characterized by plenty of open fields and oaks draped with Spanish moss arching over some of its main thoroughfares; in some of the Black neighborhoods, the roads were still red clay.

The Holifields were moving to a city with a relatively strong Black middle class, a population that was better paid and educated, on average, than in the rest of the state. But that advantage only went so far to protect the Black residents from the insults of the Jim Crow South, a bastion of brutality, deprivation, and inequality all premised on the presumption of white superiority. The threat of violence reinforced the segregation of restaurants, bathrooms, and movie theaters. It was a city that still acted on anti-fraternization laws by occasionally arresting a Black person and a white person who gathered over a beer in public.

And yet Tallahassee had the benefit of harboring a rare haven: a Black institution of higher education named Florida A&M University, or as it was more commonly called, FAMU (pronounced "fam-you"). FAMU became one of the largest and best known of dozens of Black colleges founded in the years after the Civil War, especially in the South, as churches or the Freedmen's Bureau invested in schools urgently needed to help some four million newly freed people gain access to opportunity. As the Holifields grew and expanded their family, FAMU was simultaneously flourishing, its population soaring to two thousand students by 1949, achieving university status in 1953, and strengthening its graduate law and nursing programs.

The Holifields lived in a neighborhood informally known as the FAMU subdivision, just a few minutes' walk from the campus. The university was situated on the highest hill in Tallahassee, its grounds dotted with Georgian-revival red brick buildings and a neoclassical library. Built on land that once served as the plantation grounds of a territorial governor of Florida, FAMU reclaimed that landscape as one of Florida's most significant engines of growth for Black students.

In the Holifield neighborhood so close to campus, leading Black experts in politics, art, and finance mowed lawns in front of relatively new ranch homes. Millicent played bridge with neighbors such as Carriemae Marquess, a PhD who led FAMU's institutional food department (and who died in 2020). In the fifties, Millicent pursued her bachelor of science degree at FAMU while working at its university hospital, which was, at the time, the only hospital in northern Florida that treated Black patients. Her son Bishop would later deliver the high school newspaper to the homes of dozens of FAMU academics and PhDs in his neighborhood, as well as the renowned football coach Jake Gaither, who, in 1969, led FAMU's team to victory against the University of Tampa—the historic first game in which an HBCU played against a predominantly white football team in the South.

In recent years, the advantages for Black students who attend an HBCU have become apparent and widely publicized: Black students who attend those schools are more likely to report feeling supported than Black students who attend non-HBCUs; they represent a disproportionately high percentage of Black students who attend medical school or enroll in PhD programs for STEM (science, technology, engineering, and mathematics) fields; they even report better health outcomes down the road. But Black Tallahasseans who did not teach or study at FAMU also benefited from the opportunities for enrichment and the concentration of talented, like-minded people that FAMU provided; the neighborhood itself served as a kind of benevolent support for children, shaping values but also making a first-rate education as available—as conquerable—as the lake in the Groff's backyard was to Sarah Groff.

The Harvard economist Raj Chetty is famous for showing the power of a neighborhood in the United States, proving with data the significance of a zip code in shaping children's outcomes. He has found that the longer children lived in neighborhoods that were advantaged, the more education they ended up pursuing, suggesting that it was the neighborhood (as opposed to, say, the children's parents) that was shaping such an influential life choice. In related research, he and his team looked specifically at what gave children the opportunity to become innovators. In studying zip codes that produced patent holders, Chetty found that exposure was crucial. "Children who grow up in areas with more inventors—and are thereby more exposed to innovation while growing up—are much more likely to become inventors themselves," he and his colleagues wrote in a summary of their findings in 2017. In examining patent holders' elementary school math scores, as well as tax records, Chetty also found that it wasn't enough for children to be talented in that subject—only the affluent students with those strong scores seemed to have a chance at becoming a patent holder, suggesting that talent was going untapped because of inequality and a lack of proximity to opportunity, with too many "lost Einsteins," as he put it, never having the chance to explore their ideas. Chetty's work, in general, highlights the injustices and collective costs of economic inequality, but also reflects how powerful a neighborhood could be in making certain kinds of ambitious thinking possible—surely in an area like Silicon Valley, but also, possibly, a college town like FAMU.

In Tallahassee, the Holifield parents extracted all that they could from FAMU: They took their children to hear the FAMU symphonic band, or to enjoy Count Basie when he performed on campus, or to take in a production of Lorraine Hansberry's *A Raisin in the Sun*. As soon as their children were old enough, they enrolled them in FAMU's Demonstration School, which had a reputation for excellence, and was staffed by teachers with advanced degrees, many of them graduates of FAMU. "It was a cocoon," said Hansel Tookes, a friend of the Holifields who attended the same school. "When we grew up, we didn't think that there were things we couldn't do." That didn't mean they were blind to the brutal injustice outside the cocoon. "When saying the Pledge of Allegiance, we said, 'with liberty and justice—for some people,'" Tookes recalled. "We knew what was around us."

The Tookes family was one of many in the neighborhood who were renowned. Hansel Tookes went on to serve as president of Raytheon International; his brother, Darryl Tookes, is a composer and musician who recorded with Leonard Bernstein and Quincy Jones, among many others, and who has taught at the Tisch School of the Arts and at FAMU, where he runs a music industry education program. (Darryl's daughter, Ryan, recently landed a job in Gainesville, Florida—working at the bookstore recently opened by Lauren Groff.)

The Holifields' lives in Tallahassee were intertwined with the school, a protective institution where a curious, art-loving girl like Marilyn would take lessons from graduate students trained by Samella Lewis, an artist and historian who would go on to write a seminal and comprehensive textbook on Black art; where all three children participated in children's theater, trained to think about how they held their bodies, how they could make themselves heard; where they thrilled to the thundering sound of the Marching 100, FAMU's celebrated marching band; where Marilyn enjoyed a journalism workshop for high school students from around the state. That the Holifields landed in a city with an HBCU was an exceptional stroke of luck: there have never been more than roughly 120 four-year HBCUs in the country (the number reached its peak in the 1930s).

Bishop Sr.'s federal job provided him job security in Tallahassee, and the nearby school provided enrichment for his family; but for him, those were both starting places. Beyond FAMU, he saw opportunity in the availability of land in a city that was clearly growing fast. Shortly after moving to Tallahassee, Bishop Sr. started buying empty lots where he hoped to build housing, as well as a sixty-acre farm outside of town, where he raised cattle, pine trees for lumber, and grass that he sold to nurseries. In the name of saving for investments, their very home environment was designed for thrift: They bought a two-family house and rented out half to generate income. "He had a knack for making things grow," Bishop said of his father.

While the Holifields were expanding their real estate holdings, Millicent, who worked at the hospital, was also focusing on a way to gain autonomy, and to bring other Black women into the nursing profession. For several years, she petitioned the state, hoping to open a licensed practical nursing program for Black women in Tallahassee, a school that would be the first of its kind in the area. FAMU had established a nursing program for Black women to become registered nurses in 1936; but Millicent wanted to offer a more accessible one-year program equivalent to those available to young white women— a fast track into the workforce open to the working poor, a credential that was within reach and that could nonetheless catapult those women into the middle class, as nursing historically has done for so many women. In 1957, after at least three years of appeals, the state finally said yes to Millicent's nursing school for Black women. "That was a happy day," Marilyn recalled. "That was joy. We celebrated."

That October, Millicent Holifield launched the city's first licensed practical nursing program open to Black students, with a class of twenty; only thirteen were able to uphold her standards and graduate. The city had carved out a space for her in "about the ricketiest building they could find" on the premises of what was then the city's largest high school for Black students, she told *The Tallahassee Democrat* many years later. One of her students in that first graduating group had been working at a local five-and-dime; another had dropped out of school after eighth grade and was picking tobacco.

Over the years, Millicent kept pushing—usually with success—for improvements. The books had to be unused; the students had to get a stipend; the lab had to be sparkling new. Millicent was "a petite, brilliant woman with the go-get of a pit bull," as her former student Janice Stanley described her.

At the end of her first year, when it was time for her students to take the state nursing test, Millicent would not stand for anyone in her class to fail. If a student did not show up for class, Millicent had no qualms about showing up at her home, marching into her bedroom, and hustling her out of bed. The students she retained had a near-perfect pass rate for the test.

That her children excelled, that her students excelled, was not a matter of vanity but was an imperative she deemed essential for their future prospects, she made clear. "I trained my nurses like I trained my children," she told *The Tallahassee Democrat* in 1990, in an article that reveals much about the governing principles of the Holifield household. "I exposed my girls to more than most of the white nursing students got because that was the only way they had a chance—to be the very best."

And the Holifield children did excel, living up to expectations that were explicit and high, the goal always clear: "We had to do better than they did," Bishop said. His father braved a dangerous journey as a child and lived far from home just to get an education; all his own children had to do, Bishop Sr. often pointed out, was show up for class.

As with her students, Millicent left little up to chance with her children. "My mother was the one with the plan," said Bishop. "It was our job to execute it." The Holifield siblings all attended summer school, not the kind for children who had fallen behind but for those who wanted to jump ahead—or whose parents wanted them to jump ahead. When they went to the library, Millicent expected them to check out two nonfiction books plus two works of fiction. They were all expected to bring home superb grades. They would all attend college and receive professional or graduate degrees, ideally studying at some point up north, where their parents considered the education to be superior and the opportunities better for Black young people. And when they graduated, they were expected to earn a good living. "A Black person needed financial independence so that they wouldn't have to lose their self-respect dealing with white people," said Marilyn, paraphrasing her father's thinking at the time. At a time when the very law of the land insisted on the premise of the inferiority of Black people, the Holifield parents assertively counterprogrammed, following family tradition. Bishop Sr.'s father—grandfather to Bishop Jr., Ed, and Marilyn—never let any of his children work for a white man back in Mississippi, according to Bishop Sr.'s cousin Barbara Rhodes. Bishop Sr., as a father himself, would not permit his children to attend a movie theater if the seating was segregated, and they were not allowed to watch *Amos 'n' Andy,* a show that relied on stereotypes of Black people for its humor.

In 1962, the oldest Holifield child, Bishop, chose to enroll in FAMU, having finished high school after three years, too young to leave home, but too advanced to continue at his school. The environment at FAMU was already charged with the energy of resistance and change: By then, FAMU had become the locus of one of the earliest and most successful bus boycotts in the country. In 1960, two FAMU graduates, sisters Priscilla and Patricia Stephens, were among the first activists to choose to serve their jail time rather than accept bail following desegregation efforts. The strategic breakthrough helped elevate the cause of nonviolent direct action; the Rev. Dr. Martin Luther King, Jr., wrote to Patricia in a telegram that she would "bring all of America [to] the threshold of the world's bright tomorrows."

While Bishop was thriving at FAMU, his sister Marilyn was enduring an entirely different educational experience, one that was as alienating as Bishop's college years were affirming. In 1963, the Holifield parents were faced with one of the hazards of raising children to be ambitious: the possibility that the drive and values parents have fostered in their children will take on a force of their own. They planted seeds that developed in a climate that was unlike the one in which they came of age, and were inevitably surprised, perhaps even unnerved, by the unexpected ways those seeds developed.

In ninth grade, in 1961, Marilyn came to her parents with a proposal she fully expected them to endorse: An elite boarding school up north had recently accepted a young man she knew down the street, and now Marilyn had decided she wanted to attend a similar school that accepted girls. She appreciated the teachers at her own school, but she was aware of the obvious inequity of the resources at their disposal. The textbooks were handed down, inscribed with the names of white students who had completed high school elsewhere many years earlier. Trying to study a slide in a science class meant waiting in line for the quickest of glimpses, as thirty or so students shared one microscope.

She went ahead and filled out the application, confident that her parents would endorse this decision. They had always told her the schools up north would be better for Black students, and they clearly valued excellence in education, whatever the cost.

I can imagine that the Holifields might have felt conflicted, appreciating how their own personal histories might inspire Marilyn, while wondering why, when she had other good options, she would want to endure any of what they had: Millicent, studying nursing as the rare Black student among so many white ones, and Bishop Sr., traveling so far from home at that young age to obtain a high school degree.

Marilyn was fully prepared for them to say yes. Instead, the answer came back no. She was too young to leave home, they said.

"I was very, very disappointed," Marilyn said.

A year after her parents said no to boarding school up north, Marilyn, still intent on pursuing educational excellence, went to them with a counterproposal that was, in some ways, even more radical: She wanted to attend Leon High School.

Leon High, then considered the best high school in Tallahassee and one of the best in Florida, was a dramatically situated city landmark, recalled by its alumni as "the grand old school on a hill." Proud and historic, Leon High educated the children of the state's governors, including the daughters of Cecil Bryant, a segregationist who promised shortly after he was elected in 1960 to "maintain segregation without violence, anarchy, or closing of schools." The school was almost entirely white, with strong Southern roots—well into the late sixties, it continued to hold a Confederate Ball (the theme in 1968: "Everything's Coming Up Dixie").

In 1954, the US Supreme Court decision in *Brown v. Board of Education* had held that school segregation was unconstitutional, but it took until 1963 for the Leon County School Board to submit a desegregation plan that was approved by a federal court. In the fall of Marilyn's junior year, for the first time, Black children would attend Leon High. Marilyn's parents agreed that she could be one of them.

Many of the young people who desegregated schools were the children of activists in the NAACP or other civil rights groups— children whose parents chose for them the path of resistance. Marilyn's parents, by contrast, would have preferred that she continue to attend the FAMU Demonstration School, she sensed. But Marilyn had been raised to expect the best; she couldn't understand why she shouldn't pursue it. "So many had suffered to open doors, and I wanted to go through them and see what was on the other side," Marilyn said.

By then, her parents would have been aware of the depredations so many Black students endured as the first to desegregate various high schools around the country: the jeers, the punches, the spitting, the isolation. To some degree, Marilyn knew, too. But somehow, she said, with the blindness of youth, or the confidence, "I didn't think it would happen to me." Marilyn did not see herself as particularly brave or unusual, and she assumed there would be many Black students who felt the way she did—eager to enroll at Leon, to get access to the best school in town. It came as a shock when she learned she would be one of only three Black students in a school of two thousand. Bishop, who by then was a sophomore at FAMU, had the opposite reaction: "I was surprised that many showed up."

At the FAMU-affiliated high school she had attended with her brothers, Marilyn had been recognized as a beauty, a star student; she was chatty, a gifted piano player, a girl with close girlfriends and closely guarded crushes. But at Leon High, the external reinforcement of who she knew herself to be was replaced by what she described as "immense isolation." The administration of Leon High had told parents of the three Black students who enrolled that their children would not attend gym class or play contact sports, for their own safety, nor could the administrators guarantee them protection in the bathrooms. Every day at lunch, Marilyn sat with Phillip Hadley, one of the other two Black students, or alone (the schedule of Harold Knowles, the third student, did not align with theirs). When no empty table was available at the cafeteria, she and Hadley knew that any students they sat next to would instantly scatter.

Every day, all day long, Marilyn heard the N-word in the halls and the cafeteria. Sometimes it was "brainy [N-word]." "I did not know that it would be up-close-and-personal hatred," she said. "You read about things, you see things on TV, you understand how mean and hateful people are. That is very different from a close-up encounter that allows you to come face-to-face, nose-to-nose, bookbag-to-bookbag, with hate." Often, Marilyn speaks slowly and carefully about that time in her life; I could almost sense the wariness she felt in searching her memory, as if she were digging among possible explosives.

About three months into the first year she spent at Leon High, Marilyn exited the bus one morning to find a gang of kids waiting and calling out names. That was not unusual, but this time they threw eggs at the three young people who were making history at Leon High. Marilyn, for some reason, was the only one who was hit. She stood there, stunned and dripping, until a janitor—the only Black adult at the school, in her recollection—drove her home.

Until we spoke of it, she said, she had not shared that story with anyone outside her family. Silence, too, was something she felt she learned from her parents, for better or for worse: not to dwell on the most painful experiences, maybe not even to talk about them.

Both her parents worked outside the home, but her father had managed to be there when she arrived at the house. She remembered her mother arriving soon after, and that she was angrier than her father. She recalled, with some emotion, that "the three of us hugged." The unspoken message was "*Marilyn, we're here for you and we support you,*" she said. "I always knew if I wanted to leave Leon High, I could. I never felt I would have disappointed them if I hadn't kept going. But I would have considered it a failure."

Maybe Marilyn's parents said yes to a choice that they knew risked so much because they knew their daughter well—they may have thought it was the right thing to do, but they perhaps also believed it was the right thing to do for her, in particular. Maybe they feared that holding Marilyn back from a bold choice that she desperately wanted would be at least as painful as watching her go through it and come out on the other side somehow stronger. They must have had strong faith that she would.

The challenging, at times harrowing, two years that Marilyn spent at Leon never penetrated a protective seal she had around her self-respect. She had thoroughly internalized the fierce messaging about her worth that her parents delivered freely throughout her childhood. "They may say that Black people are inferior, but you're not inferior," she said, paraphrasing what they told her. "When I encountered those bad experiences at Leon, I just felt I was more intelligent than all those other people." She had been raised to understand racism as tantamount to ignorance. "And it was their loss that they didn't get to know me. If I gave up, I was giving in to their meanness, their hatred."

Her graduation was marred by one last humiliation, when she reluctantly attended a party that another student was throwing for the class. When a white boy named John Herz asked Marilyn to dance, chaos ensued, according to Harold Knowles, who described the scene to the historian Glenda Alice Rabby. "While they were dancing," wrote Rabby in *The Pain and the Promise*, a history of the civil rights movement in Tallahassee, "the host's father stormed into the room, grabbed Herz by the collar, dragged him down the long front yard to his car, and literally threw him into the street. 'Get off my property,' he yelled, 'and don't come back.' Sickened and embarrassed, the three Black students left the party." Marilyn remembers the incident, but only vaguely—the dancing, followed by distress, commotion, students running away in all directions. "That was . . . just another Tuesday," she said.

Marilyn graduated seventh in her class, her protective shell thickened. She was quieter, but also steelier. She was wary, but not so wary that it slowed her down. "Marilyn isn't an 'I can take it' person," said one of her oldest friends, Marilyn Allman Maye, who met Marilyn Holifield in college. "She's more like, 'Bring it.'"

BISHOP, ENROLLED AT FAMU, had watched in awe and with pride as his sister chose to endure her last two years of high school in an environment of hate. "Marilyn was definitely an inspiration," said Bishop. "I saw that she was doing her part to try to make things better. I felt like I had to do the same."

Encouraged by his FAMU professors, Bishop enrolled at Harvard Law School in 1966, at which point the school had been aggressively recruiting Black students for several years—about 5 percent of its class was Black by the time he arrived. But the environment was a harsh change from the warm, encouraging campus he had left behind at FAMU in Tallahassee. Some students reported that professors rarely called on Black students in their classes; others felt the professors set out to humiliate them. And Bishop, having come from FAMU, was shocked by the total absence of Black professors who were tenured or on track for tenure. He didn't even see Black contractors working on the premises. He had expected a place where he could learn how to emulate his hero, the lawyer Clarence Darrow, who had successfully defended a Black college student, named Henry Sweet, when he faced charges of murdering a white man while defending his family in their home; instead, Bishop felt the law school seemed best suited to teach him how to restructure a major corporation in bankruptcy.

Bishop returned to Tallahassee the summer after his first year with the confidence—and the institutional support, from the Law Students Civil Rights Research Council, to work on the ballot initiative to reopen the pool. And when he returned to Harvard that fall, he arrived with an enhanced sense of what he could set in motion. Growing up, Bishop had seen how meaningful it was for his father to be a part of a Black affiliation group he had cofounded, the Federal Employees' Club, a mostly social Tallahassee-based organization that supported its members as they fought their way into the middle class or beyond, despite the roadblocks of the segregationist South. Now at Harvard Law, Bishop and a friend, Reginald Gilliam, started slowly organizing the Harvard Black Law Students Association, a space for camaraderie—they fielded a Black basketball team—but also for building power.

Affinity groups for marginalized communities are now a mainstay of campus life, but at the time, the formation of such a group, especially for Black students, was controversial and novel: The group Bishop Holifield cofounded was the first affinity group of its kind on a law school campus. The concern, said Alan Dershowitz, a professor at Harvard Law School at the time, who supported Bishop's organization, was that it might be legally problematic as a form of reverse discrimination.

The Holifield approach to reaching goals, which the children had watched their mother demonstrate in her unceasing effort to get her nursing program off the ground, can best be described as understated but persistent. Their entreaties are like so many flakes in a slow, steady snowfall. The snow keeps falling, until the view is a wall of snow. The Holifield way: to organize people and resources, then wait patiently for the moment of maximum leverage. The Harvard Black Law Students Association started drawing up a list of its objectives: more than token representation of Black students at the law school; HBLSA representation on the admissions and curriculum committee; and a curriculum that was more responsive to the needs of the disadvantaged. Instead of focusing on corporate creditors' rights, the HBLSA's founders believed, students should also be learning how to help individuals in debt. Bishop never planned to go to Wall Street or to work in a corporate firm; his goal was to return to Tallahassee with the skills to advance equality and protect civil rights.

On April 4, 1968, when Bishop learned that Martin Luther King, Jr., had been assassinated, his fellow students were stumbling around campus, stunned, embracing and weeping. In cities such as Washington, DC, Baltimore, and Chicago, riots began almost immediately. In nearby Boston, James Brown is credited with sparing the city the same fate, continuing on with a scheduled concert and calling for calm as a tribute to Martin Luther King, Jr.'s memory.

That evening in Cambridge, members of HBLSA gravitated to Bishop's small dorm room in Story Hall. "There was a lot of pain and anger," said Bishop. "But then we switched into action mode." The acting dean of the law school, A. James Casner, came the next day to Bishop's dorm, where he and Bishop, then twenty-three, met to talk across a desk in a conference room. Bishop's planning had met its moment: He presented a list of what the HBLSA wanted to see change at the school. Holifield's chief demand: that the law school hire a Black tenure-track member.

The following spring, Derek Bok, then the dean of Harvard Law School, along with a member of the HBLSA, visited Derrick Bell, then a highly regarded civil rights attorney who was teaching at the University of Southern California, in an effort to recruit him. Because of Harvard's prestige, Bell's hiring and subsequent development as an academic activist marked a turning point for law schools around the country. "Right after Bell's hiring is when you begin to see other schools hiring their first Black faculty members," said David Wilkins, who runs the Center on the Legal Profession at Harvard Law School.

The law school also started expanding its curriculum, per Bishop's list of demands. It added a class that would help future lawyers work with people facing debt, and it scrapped a major construction contract so it could be rewritten to ensure the hiring of minorities. Members of HBLSA went on to serve on many key committees at the law school, including the admissions and curriculum committees. Bishop is generally an understated speaker, someone who errs on the side of restraint. But he made it clear how personally urgent he considered the work that he did there. "If I hadn't succeeded," Bishop said, "I'm not sure I could have survived at Harvard."

SIBLINGS ARE THE UNDERLING collaborators in a given family, but they are also forever observing one another, evaluating one another's strengths or weaknesses, the way they curry favor, deflect blame, win praise. Outside the home, they observe one another interacting in the wild, unobserved by parents or even adults. How do they fare in a large group? Are they leaders, likable? Parents dote, but siblings see their brothers and sisters more dispassionately, with a shrewd, evaluative eye. When Charlotte Brontë took it upon herself to read Emily's poetry—with or without her permission—she saw her and saw her potential, in a way no one else had up to that point. Years before that, Charlotte's younger brother, Branwell, made (somewhat desperate) appeals to a magazine where he hoped to publish and to his favorite writers for feedback on his poetry, which may have inspired the initiative that Charlotte took, at age twenty, to see what poet laureate Robert Southey thought of her own poetry, a venture that yielded a patronizing response but that nonetheless reflected her seriousness of purpose.

Teenagers, who are just old enough to make decisions of some consequence to their lives, are temperamentally disinclined to accept a parent's advice; they are more likely to be receptive to suggestions from a caring sibling. Over the course of reporting this book, I have thought often about my own older brother, home from college, standing in the doorway of my bedroom, strongly encouraging me—I might even say bullying me—to start a student newspaper at my high school, which had none. Would I be shirking my civic responsibility, as he insisted I would be, should I fail to rise to the occasion? Worse, would I let my older brother down? In short order, I took his advice, found an adviser, cajoled fellow students. I am sure I did not advance local democracy, but I did find my vocation. Had my parents come on equally strongly with the suggestion, I would have rejected the idea altogether. I never could have felt ownership of it.

A sea of scholarship has been devoted to the influence parents have on their children, research that dwarfs the body of work dedicated to understanding how siblings shape one another. And yet evidence has clearly emerged supporting the notion that siblings do affect one another's life choices, especially their education, in significant ways. For example, especially in families in which college attendance is not a given, an older sibling's choice of a high-quality college dramatically increases the likelihood that younger siblings will apply to the same college or another that has strong graduation rates, researchers reported in a working paper published by the National Bureau of Economic Research in 2019. (The study looked at older siblings whose scores were similar—some were lucky enough to gain acceptance, some were not, a random outcome that nonetheless had an important impact on their younger siblings.)

For the Holifields, college was a given. But, even so, some of the most momentous choices in Marilyn Holifield's life happened because one of her brothers had a vision of what she could do and challenged her to do it. When Marilyn was a senior at Leon, Ed, her older brother, was attending Franklin & Marshall College in Pennsylvania, enduring a first year at a school that had few Black students, and where he was so lonely that he would have come home if he thought his parents would abide it. At college, he told a friend his younger sister was so smart, she could probably be accepted at a nearby college that was even more prestigious than his own: Swarthmore. *You should apply,* he told Marilyn, who'd never heard of it, as much because he wanted to be proven right as because he thought she would genuinely enjoy it.

He was right that she would get in: Marilyn was accepted as a freshman in 1965. He was wrong about whether she would enjoy it. She had come to the North expecting a somewhat idealized environment compared to Leon High. The environment of this new, mostly white space was not nearly as harsh as Leon High, but was nonetheless painful, her feelings of alienation sparked by more subtle racial dynamics. She spent four years, she said, sensing that she and her fellow Black students were there to help white students learn about racism. Marilyn, who was a strong swimmer after all those years of racing against boys and winning at Robinson-Trueblood, was not given a spot on the swim team, supposedly because she could not dive, a reason she never fully accepted. The only Black people on campus, other than the students, were custodial staff, who were also the only adults on campus addressed by their first names.

Away from home, in a new mostly white environment, but where she found Black friends with whom to compare notes, Marilyn, who was so close to her parents, started asserting herself. The first significant fight she ever had with her mother happened when she came home for vacation having swapped out her comb-straightened hair for a natural Afro. "I thought it was beautiful," Marilyn recalls. Even Marcus Garvey had called on Black women to embrace their natural hair, but this shift was a bridge too far for Millicent. She was furious; the two fought, and Millicent stormed off, driving away and getting into an accident that landed her in the hospital. Marilyn was so distraught that when she arrived at the hospital to visit and found her mother unconscious, Marilyn fainted, overcome with guilt.

Sit-ins and protests were happening nationwide at the time, with some turning violent: At the University of Wisconsin, the police used riot sticks on students, and at Columbia, the protests left about 150 students injured at the hands of the police; an officer was permanently paralyzed after a student fell from a second-story window onto his back. In 1969, Marilyn was one of eight students who helped organize a historic sit-in at Swarthmore's admissions office, calling for more Black studies and protesting a public report on Black students at the school that the students believed violated their privacy and dignity. The protest at Swarthmore, contentious but peaceful, nonetheless ended with a devastating trauma. On the eighth morning of the sit-in, the beloved president of the campus, Courtney Smith, arrived at his office, complained to his secretary of chest pains, and within minutes died of a heart attack. His family reported that he had suffered for years from coronary disease, but an editorial in *The Philadelphia Inquirer* characterized Marilyn and her fellow activists as "arrogant little militants" in an editorial with a title that implied their guilt: "The Lesson of Courtney Smith's Death." When one of the activists emerged from the building to announce the end of the sit-in, a student in the crowd screamed out, "Murderer!" and another yelled, "You killed Courtney Smith!" Sympathetic community activists were so worried for the students' physical safety that they whisked them off campus in vans. They spent five days at a nearby church, sleeping in the pews or church offices, before the environment was considered safe for their return.

Marilyn spent the remainder of her last year at Swarthmore feeling ready to move on, yet somewhat unmoored. She had majored in economics on the advice of her brother Ed, who went on to pursue his master's degree in economics, but the material did not excite her. When Bishop spoke to Marilyn, he sensed that she was adrift and stepped in. He had been raised to take seriously his responsibility as the oldest. The one time his father had ever used corporal punishment on him was when he failed to accompany his younger siblings home from school as had been expected. Bishop considered Marilyn's talent for writing, her eloquence, her sharp mind, and resolved that she should go to Harvard Law School. If he had managed there at a less hospitable time in its history for Black students, he could only imagine how much Marilyn would gain from the experience. Besides, she had already survived Leon High—she could surely handle Harvard Law.

First, he turned his persuasive skills on Marilyn, insisting that she would thrive there. Then he turned his persuasive skills on the law school's admissions officers, with whom he worked closely as the cofounder of the HBLSA. In the fall of 1969, Marilyn followed Bishop's lead into Harvard Law School. "I'm a legacy admission," she told me, only half in jest.

In the Holifield family—and others I interviewed for this book— the parents shaped the values and expectations; then the siblings took over in wielding their influence on one another; and then, finally, they all individuated, even as they continued to help one another, with one's power and connections amplifying the others'. Marilyn and Bishop both obtained law degrees, but used them in dramatically different ways. Not long after graduating from law school, Marilyn worked for the NAACP Legal Defense and Educational Fund in New York, for whom she litigated various class action and prison reform lawsuits. But in her early thirties, she decided, for some of the same reasons she wanted to desegregate Leon, that she wanted to work at one of Florida's exclusive white-shoe law firms. At that time, those firms were still largely closed to Black lawyers; an old childhood friend all but scoffed when Marilyn told him of her specific goal. She'd never get a job, he told her, because no powerful firm in the South wanted an overeducated Black woman.

But Marilyn had been raised in a family whose unspoken motto, as she characterized it, was "all things possible." There was was no door she did not have the right to open. "It was like I wanted to step foot on Mars and prove a point—that Mars wasn't really that big of a deal," she said. She spent eighteen months job searching, eighteen months of living at home, managing Bishop's worries and keeping the faith that it would happen. Finally, once again, her older brother helped clear a path: Bishop passed on word of Marilyn's interest to an acquaintance at Holland & Knight, having heard that a fellow Black lawyer had been offered a job there but turned it down. Soon afterwards, Marilyn landed the job. In 1981, she became the first Black lawyer hired by Holland & Knight and joined its office in Tampa.

Having reached a goal she set for herself, once again, Marilyn found the experience of living it out to be emotionally trying. At times, she felt a profound sense of isolation. "I think people were reluctant to interact with me," she told *The Miami Times* in 2009. "People really did not know how to find a place for me." Eventually, she moved to the firm's office in Miami, a more cosmopolitan city where she felt more at ease and her practice thrived. In 1986, she became the first Black female law partner at any major firm in Florida.

Marilyn believes that her academic credentials were crucial in her landing that job. "One of the luxuries of going to Harvard Law School is it means you can take risks in your life," Barack Obama told a *Los Angeles Times* reporter in 1990, when he became the first Black student to edit the *Harvard Law Review*. "You can try to do things to improve society and still land on your feet. That's what a Harvard education should buy—enough confidence and security to pursue your dreams and give something back."

Marilyn offered a blunter assessment: "It reduces some of the penalties of being Black," she said. It gave her and Bishop tools, but it also helped them gain actual power.

By 1990, Marilyn was deeply entrenched in Miami. She married a developer (a marriage that ended twenty-five years later), took up sailing, joined an opera society, defended major media companies. She also served as the chairman of the board of a nonprofit addiction treatment center and ran seminars for lawyers and judges on the legal needs of the city's most disadvantaged populations. The subject of occasional profiles in *The Miami Herald*, she graduated in 1984 from a leadership program intended to foster future civic leaders in the city.

In 1990, Marilyn, then forty-two, found herself in a position to make maximum use of the experience she had accumulated and the credibility she'd earned within the business community. She was thriving, even if that success was extremely hard won, having started, in a way, with skills she built during her time at Leon; now she would use her social capital to shape the city, to try to make it more equitable for its Black population. At the time Miami was a booming city, but it struggled with dramatic economic inequality and increasing violence, unrest, and tension between the Black population and the growing Cuban American population. Resentments deepened in the fall of that year, when Nelson Mandela, the South African anti-apartheid leader who had been imprisoned for 27 years for his activism, was scheduled to speak in Miami as part of a celebratory world tour following his release. Before he arrived in Miami, Mandela, who had expressed gratitude in the past to Fidel Castro for his support, once again acknowledged his solidarity with the tyrannical leader whom so many members of Miami's large Cuban population had fled. The city's Cuban American mayor, Xavier Suarez, criticized Mandela and the city rescinded its formal welcome of Mandela—a proclamation and a key to the city—outraging the city's Black community, which already felt increasingly marginalized.

The fury around the city's treatment of Mandela coincided with news stories about reports of police brutality toward Haitian Americans in Miami, a series of events that provoked a conversation among a group of Black lawyers, including Marilyn, who had gathered to help plan a convention for an association of Black lawyers in Miami later that year. Given what was happening in Miami, they asked each other, why should they bring Miami their business? Why, for that matter, would any Black organization bring their convention there? In the months to come, Marilyn became a leading figure in a peaceful but powerful movement that would jolt the city and shift long-standing power dynamics: a tourism boycott of Miami.

Marilyn had strong memories of the success of the Tallahassee bus boycott that was organized when she was in grade school. Withholding money, even in her own home, was always a form of resistance. Her parents had never allowed her and her brothers to patronize businesses that treated Black customers as second-class citizens. If a restaurant served them food out of a separate window, the Holifields never frequented it. "My parents said, 'If we can't be there as full-fledged citizens? No part of it,'" recalled Marilyn.

The movement, known as Boycott Miami, lasted three years, as Black associations, the ACLU, and associations with large Black memberships canceled their plans to hold conventions in Miami or refrained from making plans to come there, costing the city an estimated $20 million to $50 million in lost tourism revenue. After three years of back-and-forth with business leaders in Miami, on the last night of negotiations, it was just Marilyn and the public face of the movement, H. T. Smith, an American trial lawyer who'd been active in the anti-apartheid movement, fighting to nail down every financial incentive they'd asked for on behalf of the Black community: significant resources and structural changes devoted to increasing Black participation in the tourism industry; more scholarships for Black students interested in management and hospitality; and, in what they hoped would be a visible symbol on the Miami skyline, a Black-owned premier hotel. Called the "quiet riot," the Miami boycott has been recognized as a movement that not only saw real economic victories for Black people in Miami but ushered in an ongoing conversation about the city's responsibilities to some of its most disenfranchised residents. "Over time, the city's business leaders stopped focusing on the punishment being inflicted by lost business and embraced—slowly at first, then deeply and sincerely—the Black community's frustration over its economic plight," reported *The Miami Herald* in the spring of 1993, in an article that highlighted Marilyn's leadership role in the boycott.

The "quiet riot" was a name that could equally apply to the Holifields themselves. And as was true of the Holifields, it was highly effective. Marilyn was the official co-spokesperson, but also a valued negotiator, someone who could somehow, from her perch at a corporate law firm, nonetheless be a major architect in a fight that threatened the city's revenue and reputation.

"We had a lot of crises—when running a boycott, you've got the business community against you, the politicians against you. You have no money, no nothing, except the will and the moral authority, and that was all we had," said Smith, who had previously run Free South Africa, a US-based anti-apartheid group. "And she was right there, like a tree planted by the river. She would not be moved."

Bishop also continued his activism after Harvard Law School, eventually returning home to Tallahassee to serve as general counsel at FAMU and to continue to preserve the power of the historically Black university that meant so much to his family.

After the Civil Rights Act had passed in 1964, the state argued that it no longer needed to fund two law schools, one for Black students and one for whites, and essentially transferred FAMU's law school to Florida State University. Bishop devoted close to fifteen years finding angles, arguments, and loopholes to convince the state to reopen the FAMU law school, which he saw as a crucial way to address the chronic underrepresentation of Black lawyers in the Florida bar. In 2000, the state legislature voted to reinstate the underdog school in Orlando. "I celebrate every time I think about it," said Bishop, who considers it his greatest success, and who continues to fight to strengthen the institution, even as it suffers setbacks.

Ed Holifield, the family's middle sibling, took a different path and points to the way that temperament, as much as values or intelligence, can shape such different lives among the siblings from one family. After his training in economics, he pursued a medical degree, following his mother into the field of health. As a cardiologist, he joined his brother at FAMU, where he became director of student health services. But at FAMU, the institution so beloved by his family, he became embroiled in litigation after his contract was not renewed, claiming that was done in retaliation for his calling out of unsafe conditions. (He reached a settlement with the university.) At a federal prison where he oversaw health care services, he filed another lawsuit, asserting that he had been terminated because of racial discrimination. (The suit was dismissed.) Ed has the brilliance of his siblings; his energy for a fight took a different form. He is known as an unrelenting, even contentious figure in Tallahassee, with sometimes unpopular views. In 2008, he was part of a group of activists who protested against plans for a biomass plant that would be located in a poor, predominantly Black neighborhood in Tallahassee. The company withdrew its plans. In 2021, researchers from the Harvard T. H. Chan School of Public Health released a report emphasizing that biomass was "not a clean or healthy alternative energy source."

"Of all three of us, he's done the most to save actual lives," Marilyn said. She had in mind, specifically, the focus Ed has maintained over the years on protecting the health of some of Tallahassee's vulnerable populations. In 2014, when Ed realized that Tallahassee Memorial Hospital was dispensing formula to every new mother who left the hospital, he launched a prolonged campaign to change that policy, given the body of research that links breastfeeding to better health outcomes for children. He spoke before the city and county commissions; he harassed—his word choice, used proudly—the hospital's CEO. "Basically, I embarrassed them to death," said Ed. In 1994, the hospital started strongly encouraging breastfeeding, moving toward an official "Baby-Friendly" certification by the World Health Organization. A hospital administrator overseeing the move called Holifield a "catalyst" in *The Tallahassee Democrat*. He has been credited in the press for keeping public attention—through service on a Leon County Health Advisory board, and in fiery testimony he provided to the Leon County Commission—focused on the issue of Black infant mortality rates in Tallahassee.

The Holifields rarely collaborated professionally, but they did rally together to help their parents, who, over the course of their lives, built and managed hundreds of apartment units, especially for student housing, in Tallahassee. In their later years, subject to complaints and a lawsuit about the standards maintained at some of their rental properties, they turned to Bishop and Marilyn, who stepped in to help manage the case and right the ship going forward. In the last years of their parents' lives, Marilyn and Ed were intimately involved in their care. And the siblings continued to support one another when called upon.

Bishop, his vision impaired with the recurrence of a childhood problem, nearly turned down an invitation in 2016 to be honored at a weekend-long celebration of the fiftieth anniversary of the Harvard Black Law Students Association. Marilyn turned her gift for persuasive argument on her brother, insisting that he accept, working with him closely on the speech, checking his facts, refining his language, until he essentially had it memorized. When the crowd rose to give him a standing ovation, she and Ed were among them. Bishop, in turn, helped lobby for Marilyn's election, in 2018, to the Harvard Board of Overseers (a role from which she is now retired). Robert Bell, a friend of Bishop's from the HBLSA who went on to serve as chief judge on the Maryland Court of Appeals, received a phone call from Bishop, advocating on his sister's behalf, still bragging about his sister, still pulling for her after all those years.

MARILYN WAS ALWAYS CAREFUL about how she hoped her parents would be characterized in context. She admired them, and thought they were worthy of recognition, while also resisting the idea of casting them as "exceptional"—as if they were the rare Black family in Tallahassee to do remarkable work, when, in fact, as would be true of any community, they were one of many.

At the same time, she and her siblings are proud enough of their own family history that they have fought to make sure their parents are recognized. Ed was instrumental in the naming of a vocational school after his mother, the Millicent Holifield Academy of Health Sciences, which offers easily accessible medical and dental services at a Tallahassee high school.

Whenever I spent time with the Holifields, I was struck by the clear physical family resemblances among them and yet how different they were in sensibility, either because of the paths their lives took or perhaps because, by temperament, they were always destined for such different routes. Stylish and impeccably dressed, compact, and efficient, Marilyn sometimes ends or starts her sentences with a brisk, emphatic *So!* Bishop, tall and slow-moving, connects his thoughts with a long, protracted *Aaaaaand,* as if buying time to deliberate further. (In the last years of his career, he turned to tree farming, perhaps the ultimate expression of the family's commitment to slow growth.) Ed, genial on first impression, is both faster to find humor in a situation and quicker to express outrage.

Late in July of 2024, Marilyn joined Ed and Bishop for a family reunion, along with about forty extended family members and friends for a weekend of planned activities. On a bus tour of the city, Marilyn, Bishop, and Ed had the driver pause at key moments, so they could talk to the rest of the group about what they'd experienced over the course of their years in that city. They drove by the vocational school where their mother trained nurses, and they drove by Leon High, which still looks to Marilyn, she said, like "an impenetrable fortress," even as a photo of her hangs in a prominent hallway, celebrating her legacy. They drove by FAMU and by Robinson-Trueblood, now geared toward younger families, with a double-loop slide. At Robinson-Trueblood, they spoke to the group of the pools' closings, and of the loss of the family's oldest child, Gail. Later that day, when the family convened for a picnic at a pavilion by that same lake, the three siblings made no mention to the extended family, many of whom were from out of town, of the role that the lake held in their family history.

Five hundred miles from Tallahassee, in Miami, a city that Marilyn has made her own, a color-saturated mural covers the wall of a vacant building in a historically Black neighborhood called Overtown. At its center is a young Black female toddler. Her eyes are wide and luminous, and her gaze faces directly out at the viewer. On one side of her is a couple on the brink of a kiss; to her right is a grown woman, her hand raised high in a fist of resistance and power. The mural, by the painter known as "Mojo," pays homage to the history of the law in the city where Marilyn now practices: The building was once the office of Lawson E. Thomas, the first Black judge to serve in the South since Reconstruction, presiding over a racially segregated court system called the Negro Municipal Court.

That mural, and two others in that city, exist as part of a vision for Miami that Marilyn has been championing for the past eleven years, a project that calls on the love of art she developed as a young girl in a children's art program at FAMU overseen by one of her heroes, Samella Lewis. "Art is not a luxury as many people think," Lewis famously said in a quote that Marilyn recites from memory. "It is a necessity. It documents history—it helps educate people and stores knowledge for generations to come."

Her project is the Miami Museum of Contemporary Art of the African Diaspora (Miami MoCAAD), an institution that has been building support, even as it seeks funding and the location for a permanent home. Marilyn has made a virtue of the museum's years-long fluid state, building it up as a technology-forward institution with virtual reality exhibits and a website that makes the art of local and Black artists from around the world accessible. Her fundraising and promotion of the idea in the powerful art circles of Miami have kept the momentum moving forward. She has every confidence that she will eventually find a permanent home for what's already a thriving arts institution, and by now, so do I.

Privately, Marilyn has for decades been building a world-class collection of Black art that hangs in her own Miami condominium. Above a minimalist couch is a painting by Ed Clark, an abstract work with an image, in muted greens and pinks, that could be a globe seen from far above or a rendering of a mind at work, with white rays emanating energy outward. Two paintings by the American post-war painter Sam Gilliam, one of which she commissioned, are vibrant with rich, parallel ribbons of color. Her bedroom holds a more representational work by the Southern contemporary painter Jonathan Green, of a Black woman in a full billowing skirt and sun hat, seen from the back, as she looks across a lawn toward a small home. "It reminds me of my Mississippi roots," Marilyn said.

When Marilyn gave me a tour of her home, I found it, not surprisingly, immaculate and well-composed, although Marilyn swears it's more like controlled chaos in hidden spots, an image that always seemed to me an apt metaphor for Marilyn herself. As she shared with me the history of just a few of the eighty-plus pieces that are displayed throughout the space, I was especially struck by an image right outside her bedroom door, *Family Wading*, another painting by Jonathan Green. The work seems to capture something about the Holifield family itself, which ventured into unknown waters and insisted on their right to belong there. The mother, uncharacteristically glamorous for one of Green's paintings, wears a sun hat and sunglasses, with her arm draped around her daughter's shoulder. The daughter is looking at her brother; the father is looking into the mother's eyes. The sun is bright, almost harsh; and the family is the clear focus of the image. They look confident; they look at home, almost possessive of the space around them. Above all, they look strong and fearless in the water. They do not need to show it for the viewer to intuit this about that family: They know how to swim.

FOUR

Expectations

I magine a parallel universe in which time collapsed and unlikely worlds collided—a world in which Patrick Brontë crossed paths with the Holifield siblings' parents, Millicent and Bishop Sr., and they sat down to compare notes about their respective lives. They would find nothing in common in terms of their racial identity and milieu or their specific, structural hurdles, but they would find much in common in the pain of parental grief, as both lost daughters who were tragically young. Patrick Brontë's oldest two daughters, Maria and Elizabeth, died at ages 11 and 10, respectively, after falling ill at a boarding school. Gaskell's biography of Charlotte Brontë reports that Patrick Brontë used to say he could "converse with Maria on any of the leading topics of the day as freely, and with as much pleasure, as with any adult." As for Millicent and Bishop Sr., they described Gail as a girl who was "as perfect as a little girl could be," Marilyn said. "I always heard how smart she was, how cheerful, how full of life." Bishop Sr. and Patrick both suffered loss—and they both held high expectations for the children who followed the ones they lost.

Both Bishop Holifield, Sr., and Patrick Brontë grew up on farms; in Brontë's case, a tenant farm in Northern Ireland. Patrick showed, he once wrote, "an early fondness for books," and was such a prodigy that by the age of sixteen he was running his own school. At a time when class mobility was hard-won and anti-Irish bias was strong, Patrick, with the help of a patron, left the farming life and made the unlikely leap all the way to Cambridge as a kind of scholarship student. There, he excelled. An inscription inside a copy of Homer's *Iliad* he was awarded captures his pride in his success: "My prize for having always kept in the first class at St. John's College—Cambridge— P Brontë, AM, To be retained semper." The note, a mini parable in its own right, might have been as compelling to his children as the tales of warring heroes that followed.

Brontë, whose wife died when the oldest of his six children was then seven years old, spent most of his professional life as a minister in rapidly industrializing Haworth. He was also a local author, an outspoken advocate for the poor, and a dedicated—ultimately successful— crusader for a safer water supply, in a town plagued by poor sanitation and exceptionally grim mortality rates. He valued and championed public education; he allowed two of his daughters to further their studies in Brussels.

Patrick, who had come so far, could be demanding. It was "not his nature to woo obedience," as a friend of Charlotte's wrote in a letter to Gaskell. The children of clergy are often subject to high expectations, and for Patrick's children, his own life story set those expectations even higher. That the children were inspired by his achievement seems apparent in a letter that Charlotte, at age twenty-five—six years before *Jane Eyre* was published—wrote to her aunt. She was writing to ask if her aunt would help fund a school she hoped to start with her sisters. "Papa will perhaps think it a wild and ambitious scheme," she wrote her aunt. "But who ever rose in the world without ambition? When he left Ireland to go to Cambridge University, he was as ambitious as I am now. I want us all to go on. I know we have talents, and I want them to be turned to account."

If Patrick had high expectations for Charlotte, they came to bear more clearly on her personal life than her professional one. When he saw a courtship developing between Charlotte and a humble curate who worked under him, Patrick was so infuriated that he all but drove the man out of town before finally relenting. "The feeling which had been disappointed in Papa was ambition—paternal pride—ever a restless feeling," Charlotte, on the brink of turning thirty-eight, wrote in a letter to a friend shortly before she wed. She underlined the word *ambition,* in a rare indication of frustration with her father. Three days later, she wrote to Gaskell a letter that captured how much her father's high expectations had cost her: "I could almost cry sometimes that in this important action in my life, I cannot better satisfy papa's perhaps natural pride." The family tensions were eventually resolved, but the pain in the moment was raw.

Parental expectations are a recurring motif in virtually every family narrative—the Brontës', the Holifields', mine, yours. The list of potential expectations is endless: good grades, a good job, a good marriage, handy with a wrench (or a sewing needle). When they are spelled out and explicit, expectations can provide, at a minimum, a loose framework—the outlines of an idea of how to be in the world. They can be a burden, but they can also be an internalized source of confidence and pride. Charlotte did not satisfy her father's ambition in her choice of husband, perhaps, but Patrick's evident belief in her high worth may have helped instill the unusual confidence it took to imagine herself a successful novelist.

That is our fervent, intuitive assumption, at least: that we parents can influence our children by setting up clear expectations for them. For decades, psychologists and sociologists have tried to prove it, even as parenting research faces particular challenges. Government funding for research is more comfortably directed toward schools than private spaces, and kids at schools, who gather in one spot, are more easily studied than kids in separate homes in a given region.

One school-based study from 2014 did find that high expectations, expressed the right way, could inspire young people to aim higher. The researchers' main objective was not, in fact, to study expectations, but rather to try to overcome Black students' mistrust of the schools. If those students were likely to perceive negative feedback from white teachers as a reflection of racial bias, what kind of language might build trust and overcome that hurdle?

The research team, led by David Yeager, a University of Texas at Austin psychology professor, asked teachers in three seventh-grade social studies classrooms to give copious comments on their students' essays, grade them, and then handwrite two kinds of notes, one of which would be randomly assigned to each paper before it was returned. Half of the notes read: "I am giving you these comments so you'll have feedback on your paper." The other half read, "I am giving you these comments because I have very high expectations and I know you can reach them." Then all the students were given the opportunity to revise their essays.

The students who received the encouraging notes that reflected the teachers' expectations were more likely to revise their essays. The effect was especially powerful among Black students: 64 percent of those students revised the essay, compared to only 27 percent who received the neutral feedback.

Overcoming mistrust and defensiveness is likely important whenever someone is dispensing critical feedback, but maybe especially so for adolescents. "Often," Yeager said, "we are upholding high expectations, but it's not clear to the young person why—and so we leave it open to interpretation, and they're likely to interpret what we do in the worst possible light." But offering feedback alongside more generic, positive, feel-good language did not achieve the same results. There was something about setting expectations and assuring the students they could reach them that was especially effective—provided those sentiments were offered in conjunction with the extensive notes a teacher clearly spent time on, offering a road map to improvement. What doesn't work, says Yeager: Telling kids to try harder, on their own. What does work: "I have expectations, I know you can do it—and I'm here to support your getting to the finish line."

Beyond the challenge of finding funding, parenting research may struggle with peripheral factors in the analysis that can skew the results, and that are hard to control for. Those confounds might yield findings that suggest that certain parenting behaviors are beneficial, when in fact, those behaviors are simply more common in families whose children benefit overall from being raised in privileged circumstances. Parenting studies, for logistical reasons, are also unlikely to use a technique considered a hallmark of gold standard research, which is randomization: Researchers can't randomize which children are assigned to which parents, or vice versa; and many believe that the results of some parent-child studies are further muddled by genetics. Do parents who model empathetic behavior have more empathetic children? If so, did the children learn that behavior, or is that trait, at least in some measure, an expression of that child's inherited genetic predisposition for empathy?

Other problems with confounds bedevil research into parental expectations. Maybe some parents report to researchers that they set high expectations—and, sure enough, those parents' children performed better academically. How could researchers really know whether the daily experience of having a parent with high expectations influences students over the course of years, or whether parents were setting those high expectations based on encouraging cues they were getting from their children, even from their earliest years, about their natural aptitudes?

D. A. Briley, a developmental behavioral geneticist now at the University of Illinois Urbana-Champaign, had a strong interest in studying expectations, perhaps because of their own upbringing. D. A. (like the Brontë sisters) grew up with a father who was a member of the clergy; the child of a Nashville preacher, D. A. was expected to be "above all, a good reflection on the family," as they put it.

While many studies focused on how parents' expectations influence their children, starting in 2010, D. A. and colleagues tried to tease out whether—and how—children were also influencing the way their parents tweaked those expectations over time. To do that, they looked at a large dataset of about twelve hundred identical and fraternal twins. Starting at the time of the twins' birth, and continuing through fifth grade, their parents answered a series of detailed questions designed to capture their educational expectations of their children, as well as their observations of them. How much education did they expect their children to receive? Did their children have any behavior problems? Were they meeting developmental benchmarks for reading?

By the time the children in the study were four-year-olds, the parents of the identical twins had more similar expectations for their children than did the parents of the fraternal twins. Even when the fraternal twins were very young, parents who noticed differences in behavior—differences assumed to be influenced by genetics, as fraternal twins are less genetically similar than identical ones—expressed different expectations for each child. The study went on to suggest that those distinct expectations did subtly then influence the children's academic achievement, possibly in either a virtuous or a vicious cycle. Perhaps parents chose to read more to a child of whom they had developed higher expectations or engaged more with that child's teacher.

Ultimately, D. A. found that parental expectations are a complex formulation that reflect early impressions of a child's abilities, as well as the parents' preexisting beliefs—a bidirectional process. Those tweaked expectations have the power to influence the childrens' behavior, to shape it over time, so those expectations become a self-fulfilling prophecy. "Parents come into parenting with some preexisting idea of how far their kids will go," D. A. said in an email. "It is a transactional relationship, where parents set the stage (which does seem to be important and consequential), but then parents update their expectations based on aspects of their children." Less clear is how, exactly, those parents effectively set that stage, although D. A. theorized, in their paper, that they likely "transmitted beliefs, values, and perceptions of competence." (D. A.'s research team emphasized that the influence of parental expectations held even when the analysis controlled for socioeconomic status and race and ethnicity.)

The question then arises: Could parents who don't have high expectations be persuaded to raise them? In disadvantaged communities, lower expectations may reflect a recognition of structural hurdles—for example, the high odds of financial challenges that a child might have to overcome in order to eventually attend college. In 2017, researchers nonetheless met with some success in raising expectations by having trained assistants coach low-income parents of preschoolers in how to engage in educational play with their children, while also encouraging the parents to set goals for their children and themselves and reflect on their success in reaching them. By the time the children of the parents who were coached finished kindergarten, they were found to perform better academically and show more self-motivated learning than peers randomly assigned to the control group; the researchers' data, which also reflected success in raising the parents' expectations, linked the two outcomes. A follow-up study found beneficial ripple effects that lasted through the third grade.

Ratcheting up expectations may seem, for a certain kind of ambitious parent, like one way to evoke better results, even in a child who performs reasonably well in school already. D. A. found alarming the prospect of that notion taking hold in circles that were already competitive and high-achieving. The harsh impact of that added stress, D. A. believes, could have such a negative impact that it would hardly justify whatever marginal improvements the results yielded.

Surveying the research for the news site *Quartz*, the journalist Jenny Anderson declared that the science of expectations is "maddeningly nuanced." At what point do expectations—the expectations of success—veer into aspirations, more airy-fairy hopes? And can they ever have an adverse effect?

Kou Murayama, a psychology professor at the University of Tübingen in Germany, found that parents' aspirations for their children's academic success can indeed backfire. In a five-year-long study conducted in Bavaria, Germany, and then again in a survey of data from twelve thousand children in the United States, Murayama found that when parents had a realistic understanding of their children's potential, high expectations could help improve that child's performance. But if they were fabricating aspirations that were not grounded in a knowledge of their children's actual ability, those aspirations seemed to have a deleterious effect on their performance for reasons he could only theorize about—a feeling, perhaps, in the children, of achievement anxiety or a loss of autonomy.

Parents cannot, the results suggest, simply will their children's academic achievement into being. "Much of the previous literature in psychology conveyed a simple, straightforward message to parents who want to enhance their children's academic performance—aim high for your children, and your aim will come true," wrote Murayama. "On a practical front, the current study findings highlight the danger of simply raising parental aspirations to promote children's academic achievement."

More recently, some prominent researchers have expressed serious concerns about what they perceive as a crisis of high expectations. "The pressure to conform to perfect ideals has never been greater and could be the basis for an impending public health issue," Andrew Hill, a professor of sport and exercise psychology at York St. John University, in England, told *Science Daily* in 2022. His observations were based on a meta-analysis of data encompassing some twenty thousand college students in the United States, Great Britain, and Canada. Higher parental expectations go hand in hand with increasingly critical and controlling parenting, a combination that Hill and his fellow researcher, Thomas Curran, a behavioral scientist at the London

School of Economics and Political Science, worry puts young people at higher risk of depression, anxiety, and self-harm.

Imagine that you could determine the perfect point at which to set the bar—high enough to inspire children to believe in and to push themselves, but not so high that it becomes unproductive or even a mental health hazard. Even if you could effectively convey those expectations, in a manner that a child accepted and internalized, that alone would likely not be enough to motivate the child, holds a theory that psychologists have turned to since the seventies. Expectancy-value theory, as it's known, suggests that individuals are more likely to succeed if they expect that they can reach a given goal *and* also believe it has value. They must believe that they can make the honor roll or draft an essay or determine the equation of a sine graph—but they also have to find value in doing so. That's the aspect of motivation that parents sometimes overlook, and, D. A. told me, the aspect of motivation parents might consider emphasizing more.

In 2017, a group of education and psychology researchers showed how surprisingly easy it could be for parents to convey to their children the value of academic goals—and how dramatic the results could be. The researchers were interested in seeing whether they could influence the rates at which students chose science, technology, engineering, and math (STEM) classes. If students don't take those kinds of classes in high school, they're unlikely to take more advanced versions of them in college, cutting themselves off from a wide range of careers, many in high-growth and high-paying fields that are hungry for expertise. And yet plenty of students don't opt for STEM classes in high school. The material is hard, the courses sometimes dry, and the subject matter may not seem relevant to everyday life. Parents, the researchers knew from prior work, didn't tend to have counterarguments at the ready when their kids came home from school and expressed the idea that they'd never use algebra again or wondered aloud about why they'd even bother studying geometry. "Typically, they would just say, 'I don't know' or just disagree but not explain why," said Chris Rozek, an education professor at Washington University in St. Louis.

Rozek and his fellow researchers worked with a group of about 180 families in Wisconsin, families that were part of a broader re-search project that started at the time the mothers were pregnant. When the resulting children were in tenth grade, their parents were given access to a website and were mailed colorful pamphlets that spelled out all the ways that math and science could have value—or utility—in a wide variety of fields. The website and the pamphlets suggested to the parents engaging ways to talk to their children about the relevance of STEM subjects in areas those young people might find interesting: how chefs think about the chemistry of cooking, or how hairdressers manage their business finances, how physics deter-mines a car's potential for speed, or, more immediately, how math skills could help them get the best data plan for their phone. Equally importantly, the pamphlet advised parents on how best to try to con-nect with their children—to discuss those issues, for example, at quiet times in the car, to be gentle in their approach to the topic, and to do as much listening as talking.

The results were remarkable: By the end of high school, the stu-dents in the experimental group—the ones whose parents were ran-domly assigned to receive those mailings—had ACT math and science scores that were 12 percentile points higher than those in the control group, a remarkable and rare result for an intervention. The students in the higher group had also taken more STEM classes in high school, better preparing them for the possibility of pursuing those kinds of courses in college or even pursuing STEM-specific careers. Other kinds of interventions, such as tutoring, have been shown to increase test scores—but it's rare to have dramatic results in an experiment that has such low cost and radical simplicity.

What made these findings especially striking was how much influ-ence the parent seemed to have in this dynamic. It was not a guidance counselor telling the students the value of STEM, or a cherished teacher, or a doctor in a lab coat, or an influencer on TikTok, or a best friend. Helping a child care about a given goal, translating it into something that the child might find meaningful or even just useful—that is one area in which a parent's power seems strong. Even teenagers, a category of humans who seem programmed to reject their parents' values, seemed highly open to hearing—and then acting on—their parents' explanation of why a given choice had value.

The parents were crucial in achieving the outcome—but the mechanism at play did not involve a top-down transmission of pressure. The study did not focus on a quality or interest that parents wished their children had; instead, it helped parents understand how they could engage their children by invoking the visions the children themselves already cared about the most.

I thought about that dynamic in some of my later conversations with the researcher D. A. about how their own upbringing might have influenced their research, specifically their curiosity about expectations. At first, D. A. was the rare researcher who had little to say on the subject. But in between our first interview and our last, a lot had changed. In one of those later interviews, D. A. opened up a bit more about their early years: how they'd been bullied as a child growing up in Nashville and always felt somehow different, which is what led them to personality research in college. What was it that made them so different? It took D. A. years to recognize that "those differences other kids were intuitively picking up in childhood probably had something to do with gender rather than personality," they told me.

As the child of a preacher, D. A. internalized, for many years, expectations to excel—but also to have a conventional marriage and conform to typical gender norms, expectations they were only able to move past more recently when the pandemic caused such upheaval, giving D. A. the space to question whether they were still invested in meeting them.

Since then, D. A. has been energized by a new area of research: individual differences in gender identity development among gender nonconforming children. Expectations, D. A.'s earlier research found, can be motivating; but their own personal experience made it clear, they said, that "they can also saddle an individual's growth, when they are contrary to the values and preferences of the person bearing the weight of them."

IN NOVEMBER 1843, the headmaster of a school in Brussels wrote to the father of two of his students to thank him for raising two out-standing scholars. "You will undoubtedly learn with pleasure that your children have made extraordinary progress in all the branches of learning, and that this progress is entirely due to their love of work and their perseverance," he wrote. "With pupils like this we had very little to do; their progress is more your work than ours. We did not have to teach them the value of time and instruction, they had learnt all this in their paternal home."

The proud headmaster was Constantin Héger, a superb educator whose opinion meant a great deal to Charlotte. Both she and Emily, then ages twenty-seven and twenty-five, respectively, understood the value of the work they were doing; they also had likely internalized some very high expectations regarding at least their academic perfor-mance. But their father's hope had never been that they would be-come writers. The point of their education was to give them the means to support themselves, as educators of young children.

Within the Brontë family, for many years, the highest collective ambitions rested primarily on its only son, Branwell. Patrick person-ally tutored him in Latin and Greek and in the study of Virgil, Homer, and Horace; he also indicated in a letter that he intended to send his son to the prestigious Royal Academy for Artists in London. That plan never came to fruition, as it turned out, possibly for financial reasons; but even the practical training Branwell received to be a por-trait artist also ran aground. A series of jobs went downhill from there, including one as a tutor that ended when he had an affair with the mother of the child under his instruction. Back home in disgrace, he started succumbing to addiction—alcohol and quite likely opiates—before dying of tuberculosis at age thirty-one.

A month after his death, in a letter to her London publisher, Charlotte reflected on how much it pained her family that Branwell had failed to meet their high expectations. "Branwell was his Father's and his sisters' pride and hope in boyhood, but since manhood, the case has been otherwise," she wrote. "It has been our lot to see him take a wrong bent; to hope, expect, wait his return to the right path; to know the sickness of hope deferred, the dismay of prayer baffled, to experience despair at last; and now to behold the suddenly early obscure close of what might have been a noble career." She wept not for the loss of a close companion—the brother she knew and loved was long gone, taken by addiction—but "for the wreck of talent, the ruin of promise, the untimely, dreary extinction of what might have been a burning and a shining light." The measure of his artistic and professional success was not swept aside as irrelevant in the face of tragedy, as is so often the case; it was, at that moment, for Charlotte, a paramount source of grief.

The Brontë sisters had always written, with great passion, in their childhoods, and then in rare moments of leisure as adults. But it was only when the family had all but lost faith in Branwell—when he had clearly succumbed to addiction—that Charlotte finally summoned her sisters to focus on submitting their work to various publishers. Her plans to run a school with Emily and Anne hit too many stumbling blocks, she had been an unenthusiastic governess at best, and the man she loved—the same headmaster who wrote to her father, singing her praises—was married and increasingly remote. Nonetheless, at age thirty, she managed to dig deep into some well of confidence and pick up the family expectations where her brother, no longer in sound mind, had tossed them.

FIVE

The Murguías

The feminist revolution came briefly to the Murguía household in 1974, an insurrection on South Twenty-seventh Street in Kansas City, Kansas, that flared hot, then was over before it could begin to burn.

Mary, then around fourteen years old—technically the youngest in the family of seven siblings—led the charge. Mary was born just six minutes after her identical twin sister, Janet; they both grew up to be young women with sturdy shoulders, smiles that flashed, and hair that fell in dark, glossy sheets. Their personalities were more distinct: Mary was more reactive, more likely, even, to pick a real fight with her other siblings over some perceived injustice, often involving the rules of the board games over which the youngest four competed as if their futures depended on it, or at least their reputation in the family. Janet was far from reticent, but she was a shade sunnier than her twin, more even-keeled.

The siblings—at least the youngest four of them—fought with one another frequently; with their parents, Amalia and Alfredo, never. Their parents had met in a small, provincial region of central Mexico in the early 1930s, and "within the walls of our house, it was as if they never left," Mary said. Inside the house, the family spoke only Spanish; faith was everything. The girls showed their respect for their parents by deferring to their values and traditions, even if that meant they were expected, as young women, to sit down to a meal only once they had served their three older brothers.

The outside world nonetheless filtered through, at their public schools and over the radio. "*I am woman, hear me roar,*" the girls sang in unison, channeling Helen Reddy's popular anthem, "I Am Woman," as they cleared their brothers' plates from the table. "*I am strong, I am invincible,*" they sang as they tied on their traditional Mexican aprons to help their mother wash dishes. They sang it after dinner, when they ducked into the bedrooms to smooth down and fold back the sheets and blankets on their brothers' beds. Belting out that song was not an act of political resistance. Somehow, they could feel energized and empowered by the lyrics while carrying out the tasks expected of them.

And yet that evening when she was fourteen, Mary suddenly felt irate at the injustice of this division of labor. Why it happened that evening, she cannot say. But when her brother Ramón approached her at the Formica dining-room table where she was doing her homework, to remind her to make his bed, she had suddenly had it. "Ramón, you can make your own bed," she said.

Ramón objected. Mary argued back—the breakdown wasn't fair, this task made no sense. Her twin, Janet, rallied to her cause, and then the three siblings went back and forth, implicitly wrestling with the questions of tradition, duty, family obligation, and self-determination that would govern their lives for decades to come. Ramón eventually brought his father out of the bedroom to settle the matter. Mary and Janet stood side by side at one end of their long kitchen table, staring down their father and brother, who stood at the other.

The siblings recall Mary's sole uttered moment of resistance to their parents' traditions as a high-water mark for rebellion in their childhood years, a reflection of how deferent they were for so long. The respect they felt for their parents was informed by the Mexican culture of their own parents' upbringing, but it was also earned. Even as children, they appreciated the workload their mother faced daily: She did laundry for nine on a manually fed wringer washer, prepared a small restaurant's worth of meals twice daily, and watched neighbors' children for extra income. They knew that their father came home every night smelling of metal, his hands and face darkened by small shavings from the steel beams he cut at a nearby factory that was sweltering in summer, frigid in winter.

Finally, her father spoke: Make your brother's bed, he told Mary. Silently, face flushed, she did as she was told.

Two decades later, Mary had to dig deep, in the quiet fight of her life, to find the personal strength to defy, at last, her parents' expectations of what a woman could do. She is now chief judge of the U.S. Court of Appeals for the Ninth Circuit; Janet, her sister, who found her own way into realizing her aspirations, runs UnidosUS, the largest Latino civil rights and advocacy organization in the country. As a prominent female leader in the Latino community, Janet regularly gives inspiring speeches about the rights and dignity of immigrants, and Mary is also frequently asked, as a trailblazing, female Latina judge, to speak publicly about her path. Both women, when sharing their stories, credit their parents with imbuing them with the values of service and compassion that fuel their work.

Neither of the Murguía parents attended school past seventh grade; their chief ambition for their children was that they all attend college. Their mother's greatest desire for her daughters was that they stay close to home after graduating. One could argue that their daughters rose to national prominence in spite of their mother's wishes, rather than because of them. The Murguía parents did not impose expectations of great professional success; they had no connections to offer or insights into navigating channels of power.

And yet the twins' brother Ramón is also a prominent figure on the national stage, a longtime board member of the W. K. Kellogg Foundation, one of the country's largest philanthropies devoted to serving the needs of children and young people. Their older brother Carlos was a district court judge whose service at the federal level overlapped with Mary's. Both were the first Latinos to serve in their roles on their respective courts.

"My parents would have been just as proud of us today if we'd just graduated from high school," Ramón told *The Kansas City Star* in 1995. "But we keep trying to outdo one another to make our parents proud. For them, all this other stuff is unexpected; they couldn't have fathomed that."

What the siblings had going for them above all else was one another and their shared mutual project, which was to honor their parents by burnishing the Murguía name. The siblings pushed one another but also provided logistical support, connections, and counsel, comprising a valuable board of advisers with unquestionable loyalty. It would be hard to separate the interests of one from the other, as they all benefited from the story of their collective success—it all fed the "Murguía brand," as Ramón sometimes referred to it, a brand associated with triumph, upward mobility, duty, and the quest for racial equality. "We're one big chain," Carlos told *The Kansas City Star* in 2004. "If not for one, the chain would be broken." The name *Murguía* did not morph into an adjective like *Groffiness*—but if it did, the word would be synonymous with social justice.

In trying to understand the Murguías, I came to understand how pride and humility closely intermingled in their family culture. Sharing their stories with me, they felt, would be a way of honoring their parents, but also fulfilling an obligation they felt to represent some of the possibilities open to other Latinos and immigrants, as they so often did, in profiles of their family that ran in newspapers and in their own public speeches.

The connections among the Murguía siblings provided reinforcements and support as well as pressure. It was a system that offered strength when each was ascendant—but it also meant that should any of them struggle or fail, as humans inevitably do, they all would feel their foundations shake.

THE MURGUÍAS' CHILDHOOD HOME in Kansas City, Kansas, was in a part of the city known as Argentine, a heavily Mexican American neighborhood near a steel plant and the railroad where many of the families' fathers found work. Their own home was cramped and crowded but neat, with thin sheets of plastic in hot pink, yellow, and watermelon green that served as curtains. Until they left for college, five of the siblings shared a bedroom, with Mary and Janet sleeping in one full bed, Carlos and Ramón in another, while Rose Mary, their older sister, had her own twin bed. The bedroom also held one of the few nonessential items the children ever saw their father purchase— a set of encyclopedias, the other holy books in the home, stored and displayed in the top of a hutch bought expressly for that purpose. A brown towel was perpetually draped over them to protect their fronts from dust.

Their parents' bedroom was dedicated to God and was where most of the family's prayer took place outside of church. In that room, Amalia hung a crucifix and placed a rosary beside it. In May and October, the months dedicated to the rosary, the daily activities—the cooking, the basketball game, the homework, the bickering—stopped promptly at six-thirty in the evening. The whole family squeezed into the bedroom, on their knees, shoulder to shoulder, to recite that series of prayers in unison, their faith literally gathering them close. Their mother's unwavering religious devotion was so apparent that Mary sometimes felt, with some awe, that her mother had a direct line to God.

Amalia, their mother, could usually be found in the kitchen, making dinner for her own large family while simultaneously welcoming in neighbors who came by for one of her highly regarded flour tortillas or, just as often, for consolation. Mary was often mystified that her mother, whose workload was overwhelming, never seemed inclined to ask a neighbor to come back at a more convenient time; if anything, Amalia made her visitors feel she delighted in their arrival. Amalia was the oldest daughter in a family of sixteen children; she knew how to care for others. Other men ate a sandwich at the plant, but Alfredo arrived home at 12:10 P.M. every workday for a hot meal. By 12:20, when the factory whistle warning blew, he was out the door on his way back to work. "It's how he survived thirty-nine years at that plant, I'm convinced," Janet said.

How Amalia endured the labors she woke up to face every day was another question. "Don't marry," she sometimes told her daughters. Amalia had a dry wit, so that Mary and Janet knew she was not entirely serious; but they may have taken her advice to heart—of the four daughters, only one, Janet, ever married, when she was fifty-four. "It's a lot of work," Amalia used to say of family life. "I don't recommend it."

Amalia's marriage had taken her far from everyone she knew, when her husband decided they should move to Kansas City. Over the course of her life, Amalia suffered life-threatening childbirths; when Janet and Mary were toddlers, Amalia was so overwhelmed that she resorted to tying them to the legs of the kitchen table. It was true that their mother was bone-tired, exhausted, put upon; and yet her children say it was also true that she laughed—and made them laugh—as much as she sighed.

The house was too cramped to contain seven children, and the youngest four—Carlos, Ramón, Janet, and Mary—were forever jostling. But they had somewhere to go: Their house was situated in one of those neighborhoods that would prove formative, with enrichment casually convenient. There was no gleaming sports facility, no lake as in Cooperstown, no PhDs coming over to play bridge, as was the case for the Holifields. But the neighborhood was packed with immigrant families similar to the Murguías, many from the same state in Mexico. All those children spilled into the streets and found one another in the nearby park, where Carlos took charge of organizing teams. When they needed quiet—scarce to nonexistent at home—the Murguías could bring their homework to the cool, calm Carnegie library a few blocks away, where the librarians let them borrow books and also tennis rackets to be used on the park's courts. On summer days, Mary and Janet played against their brothers for hours. Their oldest brother, Alfred, taught them to bang the hell out of the ball, after he taught himself. Mary was sure that she and Janet, if they just kept at it, would get to Wimbledon eventually. "I still think we might," she told me when we met. It was Murguía humor but also Murguía ambition.

When the kids played football on a field by the railroad yard at the foot of their street—yelling out calls in English, fighting about the rules in Spanish—the train, passing nearby, rumbled so close it resonated in their bodies.

Inside the house, Mary and Janet were compliant; once outside, freed from their mother's gaze, they played football with their brothers and their friends. When the girls were about ten, a neighbor, a fellow immigrant from their region of Mexico, spied the girls midplay and informed Amalia of what she'd seen. The girls looked up from their game to see their mother coming down the street brandishing a belt and ran for their lives, hiding in an alley for hours. By the time they got home, Amalia had calmed down, but still she worried about her daughters—about the kind of women they would grow up to be, about what people back in her hometown in Mexico would say if they knew her daughters played football, that they sometimes wore pants, that, later, as young women, they stayed out past ten o'clock at night.

Amalia held her children to a high standard, maybe because she had one child about whom she worried so much that the margin of error for the others had to be small: She worried about Martha, the oldest child, a product of Amalia's first marriage, which was cut tragically short with the untimely death of her young husband. A sweet girl born with a developmental disability, Martha left school in eighth grade, and went on to work in restaurants doing kitchen prep and cleaning. All the siblings felt responsible for Martha, their big sister, and in some ways she was the family member who most shaped her siblings. She was the basis of a bond that they knew would keep them close long after each of the other siblings left home—their mother often reminded them that she would require their care when their parents were gone. Amalia's worry about Martha was never-ending, like the piles of flour tortillas she made every day. The children grew up knowing it, maybe faster because of it. They tried not to add to their parents' worries.

If Martha was the family's gravitational pull, it was Alfred, the oldest boy, who sent them all spinning out into the universe, a propulsive force, if sometimes an unpredictable one. To Janet and Mary, eight years his junior, he held the kind of glamour that only a teenager can hold for a much younger child. For nine-year-old girls who had never eaten a meal their mother had not cooked, a first trip with Alfred to Winstead's, a drive-through hamburger joint, was thrilling, a taste of pizza, a revelation. He was their bridge to the world outside the closed circuit of school and home, but even more than that, his siblings all agree, he established a clear path they could follow in the direction of opportunity.

Alfred thrived at his high school, a natural talent who competed in public speaking at the state level, who stood out on the debate team, who performed as the star in a school play, who organized a food drive and played on the varsity basketball team. Alfred expected to go to college—to be swept along on the same channel toward higher learning that carried the other strong students in his friend group.

The fall of his senior year, all his fellow honors students were asked, in alphabetical order, to attend a meeting with the guidance counselor, intended to discuss college applications. But the guidance counselors skipped Alfred altogether. Had he missed something? He sensed that his white friends—his high school was mostly white—were moving forward. Finally, after considerable time passed, he went to the guidance counselor's office. He informed the counselor that he was planning to apply to the University of Kansas, only to be told that it was now too late.

The counselor, Alfred realized too late, had assumed that Alfred would follow the same path as many other first-generation Mexican Americans from the neighborhood, young people who did not seriously consider college an option but instead went to work at the steel plant or on the Santa Fe Railway, just like their fathers. Maybe he would want that, maybe he wouldn't, Alfred told his counselor. But either way, he wanted to go to the University of Kansas.

Alfred had already won a small scholarship from the American Legion Post 213, a post that Mexican American veterans of World War II founded when another association in Kansas City refused them entry. Eventually, based on Alfred's test scores and his grades, the counselor was able to secure more funding: the American Institute of Architects could offer him a scholarship to study at the University of Kansas, if he was interested in that field. He was interested—not in architecture but in taking whatever solution presented itself. He was accepted that June, enrolled in a local community college that fall, and then, the following year, all the Kansas City Murguías crammed into a car and drove him to what everyone in Kansas City calls KU, forty-five minutes away in Lawrence, Kansas.

In attending KU, Alfred was the first in a family of siblings who would rack up many firsts of their own. None of them, including Alfred, fully appreciated how lonely it is to be a "first," or how much it would matter to his siblings to have the benefit of following in his footsteps.

ALFRED WAS ABOUT SEVEN YEARS OLD when Amalia was preg-
nant for the last time. She looked as big at five months as she had just
weeks before giving birth to her other babies. It's twins, her neighbors
told her. But when her doctor listened with his stethoscope, he heard
only one unusually strong heartbeat. "It must be a big boy," he told her.
She told her friends: "The American doctor said it's a big boy." Who
wanted to argue with the American doctor? But her neighbors had
seen Amalia pregnant before, and they knew what they knew.

During the delivery, the doctor seemed to take great pleasure in
being proven wrong. "Molly!" he cried, calling her by her American
name. "Surprise!" At the news there was a second child, she prayed;
her husband's knees nearly gave way. The doctor delivered Janet and
Mary, two beings so connected in utero that their hearts, thrumming
in time, had fooled the doctor into thinking they were one.

The twins were healthy, but after the delivery Amalia suffered a
near-fatal hemorrhage. The event was so traumatic and her recovery
so difficult that she felt compelled to visit her family in Mexico for
emotional support when the twins were still babies, a trip that she
made without her newborns, so she could rest. The twins spent several
weeks mostly separated, in the care of two different godmothers. The
first time their parents witnessed them walk was after one such separation. Reunited, the girls toddled full speed into each other's arms.

Soon after the girls were born, in September of 1960, the American doctor had taken Alfredo aside. The twins were going to suck up
a lot of attention and energy, he told him—if he was not careful, the
brother closest to them in age, Ramón, might suffer from a lack of
attention. Alfredo was not a man inclined to talk about emotions; he
was strong and hardworking, gifted at math, but not likely to agonize
over how much attention each child was getting. And yet he clearly
heard what the doctor said. With six other children to worry about,

Alfredo could not guarantee Ramón more attention, but he could give him something else—an almost twinship. In a photo taken in 1969, Janet and Mary, then about nine years old, stand beside each other on a spit of dirt just outside their home, dressed in their best: pink skater skirts and matching vests with frothy white bows beneath their chins. Ramón, then ten, stands on one side of them, and Carlos, eleven, stands on the other. The two boys are also dressed alike, in boy uniforms of a kind: khaki pants, short-sleeve white shirts, big belt buckles. Their parents deliberately dressed the boys in the same clothing, and their father was gratified when people thought he had not one but two pairs of twins—it meant he had might have successfully provided Ramón the strength of having so close an ally of his own.

In the room they shared, the four youngest siblings formed a tight unit of their own. "We were like aspen trees—people think those trees are separate, but they are really one entity," Ramón said. "We had more of a shared root system than other kids, than the other kids in our family, even. The four of us were really connected. We were more of a multicellular organism than four individuals."

In some ways the four youngest were remarkably similar, all blessed with those personality traits that psychology researchers have linked to people who thrive: They were conscientious and also open to experience—curious, eager to take advantage of whatever the environment offered, whether it was a tennis court or a museum trip with their school or the offerings of the library.

But among those four children, hierarchies were clear, and personalities distinct, even between the twins. Ramón was the quietest, a peaceable go-between; Mary was more confrontational than Janet, although both had strong wills. Carlos was the oldest of the four and the most confident of his natural role as leader, a trait that sometimes irritated Mary. The four siblings supported one another, but they also competed intensely. "Our primary way of communicating was talking smack," Mary said. They whetted their wit on one another, practiced boasting, comebacks, put-downs. Their tongues were quick; their skins were thick.

Of the four youngest, Carlos doled out the fastest and sharpest barbs. He was the most avid reader and the one who got the highest grades with the least work. He also directly benefited the most from having Alfred's mentorship. Alfred was in high school when Carlos arrived on the campus of the combined middle school and high school as a seventh grader. Alfred introduced his younger brother to his friends, showed him around the building, made him feel both protected and at home.

Following on the heels of Alfred, Carlos was intent on securing some firsts of his own in high school. "I don't know that I even wanted to be the first Latino student council president," he told me when I talked to him at Ramón's home several years ago. "I wanted to be the first Murguía." He also wanted to be the first Murguía to be valedictorian—a goal he reached, the only one of his siblings to do so. He was not striving to impress his parents; his hope was to distinguish himself amid his large, talented family, to satisfy some desire inside himself.

The Murguía parents were both natural raconteurs who typically commanded the room at large family gatherings. Their mother, especially, had long, winding, increasingly hilarious stories about her childhood—mishaps of the younger siblings she looked after, adventures she recalled from the time the circus came to her small childhood neighborhood. Carlos, too, was a natural storyteller, funny and extremely comfortable holding court. If kids wanted to know what time they were getting together to play basketball, they all knew to ask Carlos; eventually, he organized all the casual sporting events in the neighborhood, especially basketball, even founding an informal league.

During Carlos's junior year, in 1974, Alfred urged him to take advantage of an opportunity that he, to his own regret, had let pass him by: He suggested that Carlos apply for Boys State, a leadership conference, where for one week, young men chosen to represent their communities throughout the state join together to learn the basics of local, county, and state governance, forming political parties, running for office, and holding legislative sessions. Neil Armstrong went to Boys State in 1946; in 1963, a seventeen-year-old Bill Clinton attended Boys State and was then selected for the even more elite Boys Nation, where he famously joined his cohort for a visit to the White House, jostling his way to the front to shake the hand of President John F. Kennedy. Many of the civic programs' most famous alumni—Michael Jordan, Jon Bon Jovi, and Mark Wahlberg, among them—did not go into politics at all, suggesting that it selected for ambition or charisma as much as leadership skills.

Alfred's suggestion that Carlos apply was one of many small ways that Alfred shaped his younger siblings' destinies, setting them up to see themselves, even as sheltered high school students from a disadvantaged background, as future leaders capable of competing with the best in the state.

Carlos applied and was one of the boys from his high school selected to attend Boys State. Having never left his neighborhood, he showed up for the weeklong program at the University of Kansas knowing no one. Somehow, though, he had an internal confidence that propelled him to run for one of the more significant leadership positions, secretary of state. It was a strategic move, shooting for a role that was prestigious but not so desirable that the competition would be daunting. He won the election for his desired role and came home more confident than ever in his ability to succeed in the public sphere.

Ramón did almost as well in school as Carlos, enough of a natural leader that he, too, was chosen to go to Boys State. But Ramón was less outgoing than Carlos, the only one of the four youngest siblings who did not compete in debate or rhetoric. Boys State seemed to confirm for Carlos what he already knew about his leadership abilities, while for Ramón it was a rare chance to stand on his own—to make friends independent of his charismatic older brother. He ran for, and won, the humble position of mayor of his dorm.

Within their home, Mary and Janet received clear messages from their mother, messages that might have slowed them down: *A woman's place is to serve the men of her family. You will never leave the home until you marry, and maybe you should not even do that—because motherhood is a life of pain and service.* And yet they seemed capable of setting those messages aside as soon as they walked out the front door of their home. The two young women had jostled on the football field and held their own on all-neighborhood co-ed teams since they were in grade school. At school they were leaders, like their brothers. They also had the unwavering mutual support that identical twins so often share, a seamless blending of mutual investment and something close to self-interest. They were so close in age to their brothers (as was Lauren Groff to Adam) that their close identification with Carlos and Ramón might have influenced them more than the messages they were receiving from their parents about what was appropriate even to dream of.

Under those circumstances, it was only natural that, eventually, Janet would run for president of the student council and Mary would run for class president (both won). It was only natural that Janet and Mary, those star debaters, would also both try to go to Girls State, the female counterpart to Boys State.

In 1977, the year that the twins were eligible for Girls State, someone—likely Alfred—drove Janet and Mary to Shawnee, Kansas, to their interview in an upper-middle-class neighborhood of ranch houses. "They were the kind of houses where everyone had their own room," Mary remembered. Janet and Mary were strong candidates, girls who led fundraisers and co-captained the volleyball team and had a cheerful energy in their style of leadership. As a debate team, they had the benefit of being quick-witted as well as a novelty. At school, among their own classmates, they were each other's best competition: Their grades were perfectly matched, straight As in every subject but typing (they had not fully anticipated the competition posed by the girls on the secretarial track).

The day of the interview, as usual, the twins were dressed identically, confident they would both be chosen as their brothers had been in years past. Each girl went in separately to meet with a Mrs. Lickteig, a pearl-necklaced representative of the American Legion Auxiliary, the sister organization to the American Legion, the veterans' group that organized Boys State; each felt confident that she had performed well. Afterwards, Mrs. Lickteig called both girls into the room.

"You both are just tremendous, and you're actually equal," Mary recalled her saying. "I really can't make a decision between you." She explained that she did not think it would be right to send both girls, although she wished she could. She would pick one twin and someone else.

Neither Murguía twin understood why she couldn't send them both or what issue of fairness was at stake. Perhaps it seemed to this representative of the American Legion Auxiliary that in sending identical twins, she would be sending the same person twice? Later they would wonder: Was it because they had dressed alike? Was it possibly because they were both Mexican American?

Still trying to make sense of what she was saying, the girls watched Mrs. Lickteig pull out a coin. She'd resolve the dilemma with a toss, she said, and she'd do it now, in front of them, so they knew it was fair. (Reeling further, the girls questioned how this was fair.) The coin was about to leave her hand, and one of them had to make a call. "Heads," Janet said quickly.

The coin sailed upward.

BY THIS TIME IN THE GIRLS' LIVES, Carlos, three years older, was already excelling at the University of Kansas, once again benefiting from the help of his older brother. Navigating a large state school can be overwhelming for even the most mature eighteen-year-old, posing a series of bureaucratic challenges to young people still trying to adjust to being away from home for the first time. Having an older brother on campus would ease that transition.

Alfred had some of the qualities that so many older or oldest brothers in these kinds of families seem to possess—a Bishop Holifield, Jr., an Adam Groff, a Joseph Kennedy, Jr., the oldest child in the Kennedy family, on whom great hopes were pinned at birth (he died serving in World War II). Alfred was, at least in high school, demonstrably driven and confident, a good student who set high standards for his younger siblings. He also took his responsibility as the protective oldest brother seriously: As soon as Carlos arrived, Alfred found Carlos a place to live—right in Alfred's fraternity house, instantly providing the kind of social network that is especially important for a minority student. He introduced Carlos to the financial aid office staff, people with whom Alfred had formed close relationships. Especially early on, he helped him navigate a complicated system to gain access to the classes he wanted and steered him toward the better professors. "These little things actually added up to big things," said Carlos. Instead of feeling lost on a campus that size, he had a bit of an edge. He also had the comfort of being close to someone who spoke the language of home.

"Alfred, truth be told, was the key to each one of our successes in college," said Carlos, who played the same role for Ramón when Ramón followed his brothers' path to the University of Kansas. Soon after Ramón arrived, he pledged his brothers' fraternity, but he didn't make the ultimate cut. Carlos lobbied on his brother's behalf, and the fraternity brothers relented. By his senior year, Ramón had been voted the fraternity's president. The potential for him to thrive was always there—but he benefited from having an older brother who advocated for him, and an even older brother who "showed that Latino students were not to be excluded," said Ramón, who had been rejected from another fraternity, he believes, because he was Latino.

Alfred, his younger brothers thought, had shown them how they could succeed at college; and yet long after his classmates graduated, his siblings at KU were confused by his behavior. They saw that Alfred was still attending fraternity events, still sometimes showing up at undergraduate classes while he was supposedly applying to law school. Was he fulfilling some last-minute requirements? Just auditing for intellectual enrichment? Over time, Carlos and Ramón started to feel their unshakable faith in Alfred's story falter. It became increasingly clear that Alfred, their guiding star, the one who had practically invented the Murguía mythos, did not, in fact, seem to be moving toward a law degree, or acquiring any kind of a degree, not even his bachelor of arts. The siblings were dumbfounded.

In some ways, the University of Kansas had been an ideal environment for Alfred: He loved his fraternity brothers, the Jayhawks, and the encouragement of his professors. On campus, there was always another interesting class, another friend to make, another party. What he did not love, he discovered, were the classes at the architecture school. Architecture did not play to his strengths—he was highly verbal, not so much visual—and for the first time in his life he was no longer on solid academic ground. *Study psychology,* his friends suggested. Maybe he could begin again, start fresh—but what would he do with a degree in psychology? He couldn't imagine. *Then try business,* his friends told him. He worried about how he'd pay for college if he lost the scholarship tied to his study of architecture.

Alfred could still remember the home where his family lived in Kansas City before they moved to the house on Twenty-seventh Street. Three unrelated families from the same region of Mexico shared one small residence and relied on the same outhouse. The dirt outside the house, which was next door to a junkyard, was pocked with divots that filled up like small ponds when it rained. The house had no heat, so their faces were sometimes smudged with the soot that settled on them from the exhaust of a generator, fueled by kerosene, that they relied on to stay warm. Even after they moved to Twenty-seventh Street, the Murguía home had no telephone for many years; the school knew to call the neighbor next door if a problem arose. They had no books in the house, save the Bible and the encyclopedia set.

In college, as a fraternity brother, Alfred socialized with suburban, white young men whose parents were architects and doctors and lawyers and teachers and worked in finance; they took last-minute trips to Florida and spent money on restaurant meals. They seemed to understand how to succeed at college in ways that eluded Alfred. "I got up there and it was an ocean full of opportunity, but you had to know how to navigate the water," said Alfred. "Everybody there was a legacy—knew something their brother, parents, or grandparents had done. They knew how to make things happen. And I had to figure it out. Or try to."

Despite the camaraderie, he found it hard to imagine himself climbing all the steps it would take to transport himself into a life so different from the one from which he came. School started to feel like a waste of time, which was to say, a waste of money.

Money was always short at the Murguía house; Ramón estimates that his father rarely had more than $25 left over after his paycheck covered their family's basic expenses. Eager to help and adrift at school, Alfred took a job at the airport, loading and unloading luggage. The pressure he felt to succeed at school, the guilt, the feeling that he was disappointing everyone around him—making money, and giving some of it back to the family, alleviated that.

"I got a little twisted there at the end," Alfred told me, clearly still anguished by how his life unfolded during those college years. The proud child of an immigrant mother, a golden boy on whom everyone heaped expectations, he struggled to improvise some route to an ever-receding pinnacle.

Eventually, Alfred dropped out. But he deliberately misled his siblings, spending time on campus, claiming he was finishing his degree, talking about law school. "I was trying to figure it out and make it happen, and I knew what I should have been doing," he said. "Then all of the sudden, I just stopped." He always wanted to be a role model to the younger siblings who leaned on him. What if they found out that even he couldn't do all that? "I mean, the thing I was telling them to do?" he said. "I stopped doing it." If they knew, would they lose their faith not just in him but in themselves? He could not bring himself to tell the truth.

Eventually, of course, Alfred's siblings did find out. "It was a cloud, and a disappointment," said Mary. The family pieced the facts together in furtive conversations, but never raised the topic directly with Alfred. It became one of those family secrets so large that no one could get their arms around it, a mountain of ice between them that never melted.

If Alfred's experience at the University of Kansas came as a shock to his family, it is all too common for low-income students who are the first in their family to attend college: Forty percent of all students who fit that description dropped out within six years, the Pell Institute reported in 2011. College is hard, but harder still for those who feel culturally isolated or worried about money problems back home, who suffer under the weight of expectations that make both of those tensions harder to bear. Having a family member already present on campus can dramatically transform a first-generation student's experience, according to a (not yet peer-reviewed) study from 2022 that analyzed data over five years at a major unnamed university in the Northeast. "The analysis of the data revealed that as the number of family members who attended college increased, so did high school GPA, 1st semester GPA, 1st to 2nd semester retention, and 4-year graduation rate," reads the summary of the study. "Ultimately this indicates that as the number of family members who attended higher education increases so does student success. Siblings play a critical role in that they add an additional access point to social capital for the student in question."

Later, Alfred married, had three children, divorced, and went back to school to receive his college degree; he settled into a career as a banquet manager at the Kansas City Marriott, work that suits his intense sociability and his ongoing curiosity—whatever the conference, if his schedule allows, he tries to attend some of the presentations if he thinks he might learn from them. He helped raise his three children, who are now thriving adults. One recently received her PhD in the field of education at the University of Texas at Austin and is currently working as a managing director for Teach for America; another served as the legislative director and chief counsel in the office of Senator Joe Manchin; a third is a firefighter in Kansas City, Kansas.

"You embrace the opportunities that come," said Alfred. "I feel very fortunate." Often, he personally escorts the individuals at the hotel treated as VIPs, executives and elected officials who often ask after Alfred when they cross paths, socially or professionally, with one of his younger siblings.

Sometimes Alfred thinks about how much it would have helped to have someone in his life who could have told him what he always told his siblings when they considered making a leap: "Hey—go for it," he would say. "You got this."

Ramón and Carlos had each other, and both of them had Alfred; but no one else in the family—or in the school, for that matter—had the kind of unwavering, rock-solid, selfless backup that Janet and Mary had in each other. That kind of support would be a precious commodity for anyone, but maybe especially so in a family of seven children. They took all the same classes, so each had a high-achieving study partner pushing the other even further. It was as if each was twice as good as everyone else, because they each benefited from twice the intellectual firepower, twice the opportunities to pick up on a sub-tle aspect of geometry or world history that others might miss. Even their interchangeability could be an asset. One high school summer, when the owner of an ice cream store wanted to hire and train only one person, Janet accepted the job, but both girls worked there, alter-nating days, passing as one employee. The only glitch was that Mary continually failed to answer to Janet's name, a problem she solved by claiming she had a hearing problem. (This created another problem, namely, a terrible feeling of guilt from the extra solicitousness and high volume with which her boss addressed the imaginary condition when speaking to either of them.)

It was true that Janet, who was the first of the two to enter the world, tended to be first in the rare instances when it seemed to matter. Janet was first to announce that she was planning to run for president of the student council, leaving Mary to run for class president. And the year they both decided to try out for cheerleading, a neighborhood friend coached Janet on a dance routine she could use but had run out of ideas by the time Mary turned up; Janet made the team, Mary did not.

But neither of them wanted to be a cheerleader as much as they both wanted to go to Girls State. When the coin went up in the air, it was Janet who made the correct call: heads. Janet would go to the University of Kansas to have her chance to test her mettle among talented young women, to make new friends, to impress leaders of the organization. Mary would stay home.

The two girls returned to their neighborhood stunned. Mary felt envious but not resentful: They had both seen the coin toss. None of it was Janet's fault. More than anything, Mary felt a sense of injustice: this woman, imbued with power, had relied on a capricious, foolish game in a decision that affected her life. Some of the twins' friends chalked it up to the girls' being Mexican American. Mary said she felt keenly the sting of discrimination for a different reason: "Not because I was Latina, but because I was a twin," she said. What had been all her life a source of strength was finally, in this instance, what deprived her of an opportunity she knew she deserved.

Once home, her thoughts were her own. When I asked Mary if her mother or father offered her comfort and support in the face of this disappointment, she laughed out loud. "In a house that size with seven children, there was no room for feelings," she said. As much as she admired her parents, she could acknowledge that what they failed to model was open communication. No one tried to support Mary as she made alternative plans to attend a less prestigious citizenry camp that

summer; if anything, Carlos and Ramón teased her about that consolation prize, as if it were some two-bit, corny substitute.

The coin itself was so small—and yet it set off a chain of events
that led to divergent lives for the twins, emphasizing some qualities in
Janet, others in Mary. It also created their first separation, one that led
them, eventually, to lives on separate coasts.

Janet went to Girls State at the University of Kansas that summer,
along with about 550 girls from across Kansas, spending a week away
from home for the first time. Just as Carlos had, she found that her
natural leadership skills emerged. She solved the problem of too many
girls wanting to run for governor by suggesting a primary in which
the young women all voted for the three peers they considered the
strongest candidates. The group took her suggestion and rewarded her
for coming up with it by putting her on the ballot. Janet apparently
argued so persuasively during the debate that the other strongest can-
didate, Barbara Bichelmeyer, now the provost of the University of
Kansas, said she could only respond, again and again, "I have to say, I
agree with my opponent." Negotiating for power as one of the two
youngest people in a family of nine had been instructive to Janet, who
learned, at Girls State, as Ramón had before her, that she could lead
even without her closest sibling by her side.

Janet did not win the race for governor, but she was elected, the
next day, as one of two senators for Girls Nation, a weeklong event
that she attended later that summer in DC, with 101 other talented
young women from around the country. On that trip, Janet took note
of everything she experienced, wishing Mary were there to share it
with her, knowing she would tell her sister back home about everything she saw when they were reunited in Kansas City—the imposing
guard at the White House gate, the surprisingly small scale of the
building itself, the calming beauty of the Rose Garden. Janet did not
so much feel guilty as she felt the desire to share the experience as
completely as she could. "I really wanted her to catch up," said Janet.

In DC, Janet got a feel for the purposeful machinery of the office of the president, all the layers of staff laboring to advance policy to further the president's vision. The White House was no longer remote, symbolic, monolithic; it was a place that needed people to execute important work, and she could easily imagine herself as one of those people, with Mary beside her, as soon as they finished college. "It was intimidating," said Janet. "But I knew I wanted to come back to Washington, and I saw opportunities I could pursue."

After Girls Nation concluded, Janet would lie in bed beside Mary and tell her about what she had learned during that week without her. "I can tell you all the names of the people she met there," said Mary, who followed, step by step, how Janet ran her campaigns, who the competition was, how she succeeded. "I tried to figure it out, so that hopefully we weren't uneven."

After graduating from high school, the twins enrolled in KU, their family now familiar with the institution's workings. They even took at least one class alongside Ramón (whose chief goal, he said, was to land a higher grade than the twins). Eventually, Carlos, as if to make good on Alfred's unfulfilled promise, enrolled in the University of Kansas School of Law, while Ramón won a scholarship to Harvard Law School. The twins, as young women, knew their mother would never accept their leaving Kansas for any law school, so they, too, enrolled at KU's law school.

It was only after law school, when both brothers were back in Kansas City working in private practice, that Mary felt the full impact of how different she and Janet had become in at least one key regard. Girls State and Girls Nation had given Janet the experience—or maybe the confidence—to do the unthinkable: leave home.

The Murguía parents did not have visions of greatness for their children. Their focus was on survival, so much so that when Alfredo's colleagues at the plant went on strike when the girls were in high school, he refused to join them. During a previous strike he joined, he had been forced to resort to government assistance to feed his family, and he would not repeat that dependency on what he perceived as outside help. The threats of retaliatory violence were so intense that the family had a police guard outside their home; once, when the guard fell asleep, a pipe bomb exploded on the porch of their tiny home, miraculously injuring no one. It was the Murguías against the world, if it came to that.

When I talked to the Murguía siblings about their parents, I sensed an unshakable, almost religious level of respect. They loved them, but they also admired them. Their parents, both of whom had close family members who struggled with alcoholism, never touched a drink. Their friends' mothers told them their mother was unusual, exceptionally gracious, a source of strength for so many of her neighbors. And Amalia had a strong sense of her own importance as their mother. Years after Janet had left home, when she was working hard in the White House to help support Clinton's 1993 health care reform proposals, she apologized to her mother for not calling for her regular check-in, explaining that she was very busy working on behalf of the First Lady, Hillary Clinton. There was a moment of silence on the other end of the phone. Always remember, Amalia told her daughter, I am the *first* first lady.

Whenever any of the Murguía children left the house—and this continued until Alzheimer's disease took the words from Amalia—their mother would make them stand beside her for a blessing. She would raise her hand and put it on her child's head and say a prayer of benediction in Spanish: "*Que Dios te bendiga, te acompañe y que te vaya bien.*" May God bless you, be with you, and carry you safely. Each of the Murguías heard that blessing thousands of times over the course of their lives. Sometimes one of Amalia's children would already be out the door, in a rush, before she could finish, and she'd still be holding up her hand in the air, reciting the prayer over the empty space where a child had stood moments ago. "That sense of security?" said Mary. "I just think that was eternally foundational. We got the blessing of someone whose faith was so strong, you felt you could walk in front of a bus and be protected. We wouldn't leave until we heard it.

We were probably more courageous than we should have been—or maybe more unafraid is a better way to put it." Carlos also remembered the blessing as profound, "like a safety net." To Ramón, that blessing was like "a cloak of invincibility."

The blessing was a powerful symbol of how sacred the family space was, but also how threatening Amalia perceived the world outside it to be. When Ramón went away for law school, his mother did not celebrate the prestige but wept that he was moving so far away; when Carlos moved out and into his new home just a few houses away, even that called for weeping and lamentation.

Eventually, Carlos and Ramón built homes right next door to each other in the same neighborhood in which they were raised. The decision was deliberate. The houses were luxurious by the standards of the area and served many purposes. They were anchors of prosperity in the neighborhood, intended to attract investment; they were symbols that reinforced the commitment both brothers felt to the community; and they were, conveniently, less than a five-minute walk from the new home where their parents, Martha, Alfred, and Rose Mary, the second-oldest daughter, still lived, in a house across the street from the home where they grew up.

It was a coin toss that took Janet to Washington, DC, but once she got there, she was told by powerful people that she belonged there. For her, it did not require a leap of imagination to think she could leave home and build a life elsewhere. Not long after graduating from law school, she moved to DC. She was just going for an internship, she told her mother; she'd be back in a month. Then it was three months. The internship was turning into an offer; the job yielded a promotion. Every opportunity was exciting enough that her mother could not fully justify demanding her return. Janet strategically allowed her mother to get acclimated to the idea, with mixed success. "You'll barely notice I'm gone," she told her mother when she finally broke the news that she had taken a full-time job as a legislative assistant in the office of Congressman Jim Slattery. Her mother, outraged, held up her hands. "Cut off one of my fingers, and see if it doesn't hurt me," she said. Her mother, Janet said, never stopped asking her to come home—she even offered to pay her to do housework should she rejoin them.

Alfredo always wanted his children to believe that opportunity was theirs to seize in the country he and his wife had chosen as their home. "The sun shines for everyone," he used to say. "Remember, you're not better than anyone else, but you're no less than anyone else, either." And Amalia used to tell her children, "With God's help, all things are possible," an empowering sentiment as much as a spiritual one; at the same time, she clearly suffered when her children took that notion to heart, expanding their lives beyond the boundaries she imagined for them.

Janet's mother was not the only one pained by her departure. "She snuck away," said Mary. This was the great schism of their lives, the point at which their paths diverged: Once again, Janet had left without taking Mary with her. Mary took a job at the Kansas City district attorney's office, knowing that as long as she lived in town, she'd have to live at home. A woman living on her own, especially in the same city where her parents lived, was a woman estranged from her family, probably a woman with loose morals. That woman was no daughter of Amalia.

Janet thrived in Slattery's office, where her bilingual skills and warmth were invaluable resources that he relied on as he forged bonds of diplomacy. She was with him in 1985 when he flew by helicopter into the mountains of Honduras to talk to the Nicaraguan contras; she was with him in the home of Violeta Chamorro when Chamorro won the presidential election in Nicaragua in 1990 (Slattery was serving as an official observer that year). Janet traveled with him so extensively, Slattery told *The Kansas City Star* in 1990, "she was on a first-name basis with many leaders in Central America."

Over those same years, Mary developed a reputation as a fierce prosecutor who worked long hours and built her cases relentlessly; she successfully prosecuted violent murderers, child abusers, and sexual offenders. But she still could not make a case for herself: At age thirty, despite her craving for independence, she was still living at home.

She knew what her parents would say if she dared to move out: *What would they say back in our hometown in Mexico? What can you do at an apartment that you can't do at home?*

Maybe, she thought, if she got an even more prominent job, she would have the nerve to get her own apartment; but there were no openings in the US Attorney's Office in Kansas City, which would be her natural next step. Eventually, a friend of hers told her that the US Attorney's Office in Phoenix, Arizona, was hiring, and she applied for the job. She did not particularly want to leave her hometown, but she did want to live on her own, and only by landing a job outside of Kansas City—a job bigger than the one she already had—could she justify moving out. When she was offered the job of assistant United States attorney in 1990, she took it. In *The Kansas City Star,* she cited a significant pay increase as one motivation for leaving a job she had loved; but in the most roundabout, inadvertent of ways, Amalia had pushed her to make the next professional leap.

Talking about it decades later, Mary still sounded like someone who needed to justify her decision. "I didn't want to leave my family," she said, as I sat across from her in a restaurant located a few blocks from her San Francisco chambers. Recalling that pivotal moment, her voice was shaking. "But this was a good job, and I had to do it. If Janet hadn't gone to DC? I don't know if I'd have done it."

Before she told her parents, she braced herself for the scene. She told her father first, in English, the language of the world outside their family, a world that was embracing opportunities for women. Her father translated bits and pieces to her mother and reassured Mary, whose anguish he could see, that he trusted her—that Mary knew what was best for her. But her mother started shaking her head no.

"Oh, no," her mother said. "It's easy for your dad to say it." Mary knew what her mother was going to say, because she had heard it many times growing up, usually when she or Janet wanted to try something that their mother imagined to be dangerous.

"I know, Mom," Mary told her mother. "You almost died when we were born."

"Almost died?" Amalia said. "You almost killed me!"

Mary would go on to preside over the trials of domestic terrorist suspects, make agonizing judgments, go through a public confirmation hearing before the US Senate. But this story is one of the defining trials of her own life: recovering from her twin's leaving—on her worst days, it felt like an abandonment—and finding her own way forward, alone.

The Bible's story of the twins Jacob and Esau is one of the most heartrending in the scriptures, a tale of betrayal and deception that leads to parental heartbreak, a threat of murder, dueling loyalties, and the birth of warring nations. The story of their birth lends itself to competing interpretations, each equally plausible, even as one turns the other on its head. Esau was born first, the text reads, then Jacob, who "grabbed Esau's heel." Was this a combative grab or yearnful tugging? Was Jacob trying to hold his brother back so he could come out first? Or was it a gesture of love—was he trying to keep him close, to preserve, for one more minute, the sacred closeness they shared with each other and with their mother?

IN THE MID-NINETIES, on a visit to DC, Mary was walking down Constitution Avenue when, suddenly, a man she'd never met put her in a headlock, fake-pressing his knuckles into her hair. She was deeply uncomfortable but not exactly afraid: the man, she knew, was Bill Richardson, a congressman from New Mexico who went on to be the state's governor and then to run for the Democratic presidential nomination. Mary realized immediately that Richardson had mistaken her for Janet, whom Richardson apparently knew well enough to tease as if he were her big brother. Red-faced and alarmed, Mary had to explain, as politely as possible, to Richardson's great mortification, that he had the wrong person. It's one of the Murguía twins' many set-piece anecdotes, one that's maybe especially funny to people who know the twins well. Janet, as a young woman, was hardworking but temperamentally easygoing, in command of a roll-with-anything energy; Mary, by contrast, has always had the faintest shade of reserve about her, a barely there wariness, her humor a bit darker. It is impossible to imagine anyone, even Bill Richardson, even if he knew Mary well, and maybe especially if he knew her well, putting her in a headlock.

The incident, as comedic as it was, also impressed upon Mary how well-established her sister was in DC at the time, how close her relationships were in circles of power. In 1994, Janet had taken a more senior job as a special assistant to the president for legislative affairs, serving President Bill Clinton, translating the needs of one branch of the government to another, bridging worlds with the agility of an immigrant's child. Eventually promoted to deputy director for the White House Office of Legislative Affairs, she provided strategic counsel to the president from her new office in the West Wing. She was able to show her mother a photo of herself with the president and Pope John Paul II ("Okay, maybe that is a good job," her mother conceded). On one occasion, her parents had the opportunity to meet President Clinton in the Oval Office. As her mother wept with wonder, her father shook President Clinton's hand and thanked him for the opportunity he had given his daughter. "You're the ones who got her here," the president responded. The story is a hallmark of Janet's many speeches and interviews—it was personally meaningful, but also, she believes, so emblematic of the stories of so many others in the immigrant community whose parents made sacrifices so their children could contribute and thrive.

Carlos and Ramón both stayed in Kansas City. In the mid-eighties, Carlos coordinated and expanded a Kansas City program that helped undocumented immigrants there take advantage of a Reagan-era amnesty program that provided a path to citizenship. In 1990—the same year that Mary moved to Arizona—Carlos was appointed to fill a vacancy as a judge on a county district court. He successfully ran for the position in 1992, with Ramón serving as the treasurer of his political campaign.

Instead of pursuing local politics, a rough-and-tumble field for which he was not particularly well suited, Ramón channeled his energy into philanthropy, an arena in which Latinos typically had little engagement at the time. Ramón thought he could change that by tapping into a force he knew to be powerful: the Latino family. He intuited that if he asked extended families to pool donations—with each adult sibling, for example, chipping in as little as $200—the burden would feel manageable for each person, but the combined sum could be a source of collective family pride. Using that approach, he rallied his neighbors and community members to contribute to the endowment of a fund for which he was the board chair, the Greater Kansas City Hispanic Development Fund, which gave grants to nonprofits in the area that served the Latino community. When he took over the fund in 1987, it had an endowment of about $650,000; since then, he has grown that endowment to about $5 million. After building a reputation in grassroots philanthropy (he was also successful in expanding a now-flourishing scholarship fund for local Latino students), Ramón went on to serve, in 2007, as the first Latino on the board of the W. K. Kellogg Foundation. He left one well-established law firm, started his own, and married Sally Atha, the daughter of a monied Kansas City family; but he and Sally remained in Argentine, the Kansas City neighborhood where he grew up, and sent their children to the neighborhood schools. "I thought it would be difficult to advocate for neighborhoods like this one without living here," he told *The Kansas City Star* in 1998. "It certainly gives me a lot more credibility to speak as a resident, rather than a visitor."

As Ramón was receiving recognition in philanthropic circles, and

Mary was rising in the Department of Justice, eventually leading the Executive Office for US Attorneys, Janet continued to build experience and forge close bonds within the Clinton administration. Clinton, who had run on the promise to create an administration and courts that reflected the diversity of the country, made more Latino judicial nominees than any prior president, including two who came highly recommended by Janet, as well as many others: Carlos Murguía and Mary Murguía. In 1999, Carlos received Senate confirmation for a seat on the US District Court for the District of Kansas; in 2000, Mary won confirmation to the federal bench in Arizona. Both were the first Latinos in those roles.

The siblings continued to shape one another's careers, networking on one another's behalf, hatching—together—new ambitions when their futures called for new plans. For years, the youngest four siblings gathered on Sunday evenings when they were all in Kansas City, usually at the home where their parents lived, to sit around a table and talk late into the night; if they were not all in one place, they spoke on the phone. Maybe Janet could speak at one of Ramón's fundraisers; maybe someone in their vast network could use an introduction. Who in their world had recently been promoted, and how could they help someone whom they thought should be? Janet knew a board member at the Ford Foundation she thought Ramón should meet; *The Kansas City Star* would be doing another profile of the Murguía family, be prepared to take the call. Who in town needed a job, some financial support? What would be Janet's best move when Clinton left office?

In answer to that last question, Janet decided to work on Al Gore's 2000 presidential bid as a deputy campaign manager, a campaign that ended in heartbreak for Democrats. She then returned briefly to Kansas to serve as an executive vice chancellor at the University of Kansas in Lawrence. Where she would land after that campaign was yet another subject of discussion among the siblings as they sat around the kitchen table and on late-night calls. When the president of the National Council of La Raza announced that he was stepping down,

Ramón—by then an influential former board member—suggested that his sister Janet was a natural for the DC-based job, a position she accepted in 2005. (One of Janet's initiatives was to change the name of the organization to UnidosUS in 2017.)

Now one of the country's most influential civil rights leaders and advocates, Janet has overseen national campaigns that registered hundreds of thousands of Latino voters; she testifies before Congress on behalf of civil rights and economic equality, writes editorials that rally the Latino community around messaging her team has carefully crafted, and advocates for better economic opportunities and health care in the Latino community. Accustomed to fending off barbs in her childhood home, she held her own in a famously heated exchange with Lou Dobbs on CNN in 2008, accusing Dobbs, who hosted members of hate groups on his show, of "handing hate a microphone"; she was one of many voices who successfully pressured the network to remove him from the air the next year.

In July 2024, Janet was home in Kansas City with her extended family—Ramón; his wife, Sally Murguía; Rose Mary; Alfred; and Alfred's son, Nick, a firefighter in Kansas City, among others—eating fried chicken they'd all ordered in when a 202 number came in on her phone. Janet missed the call, but then a text arrived: It was Kamala Harris, calling on the day Biden endorsed her, asking for Janet's support (the UnidosUS Action Fund endorsed her the next day).

Mary, as a district court judge, rarely saw her rulings reversed and made decisions deemed important in the fields of patent and environmental law. In 2009, she oversaw a case in which Sheriff Joe Arpaio of Maricopa County was accused of racial profiling. The defendant's lawyer went on the attack, arguing that Mary could not be an impartial judge in the case because the organization her twin sister led, then called the National Council of La Raza, had strongly criticized Arpaio and his policies. In the end, Mary recused herself, to avoid even a hint of impropriety, despite her belief, as she wrote in her brief, that Arpaio's legal team was making an irrational leap about the assumed alignment of even identical twins—that "no reasonable person would automatically ascribe the views of one sibling to another."

In 2010, President Obama nominated Mary Murguía for a position on the Ninth Circuit Court of Appeals. Before questions at Mary's hearing began, Senator Dick Durbin (D-Illinois) commemorated her family for posterity in his comments on the Senate floor. "Let me just say that the story of your family is an inspiration," he said. "It is not just an immigrant story. It is an American story, and the decision of your mother and father to come to this country has made this a better nation, as we can tell from your service and the contributions which all of your family have made."

At her investiture, a ceremonial but celebratory event at which a judge is formally sworn in, Mary thanked the many family members who attended, including Carlos, pointing out that they had been the first brother and sister to sit on a federal court at the same time. "Now one of us is an appellate circuit court judge—but it's still notable," Mary said to the gathered crowd, winning a laugh at her brother's expense; it was a Murguía sibling dig she couldn't resist, even in that formal setting at a moment of her own ascension.

Alfredo died in 2001; Amalia died sixteen years later. She rarely betrayed emotion about her children's various successes, although she was often complimented on the accomplishments of her family. "Amalia, you must be so proud of your daughters," Mary and Janet heard a neighbor say to their mother one day in the kitchen. The two sisters were in the living room and stopped talking, curious to hear what their mother would say. "You know when I'll be proud?" she told her friend. "When they learn to make flour tortillas."

To Mary, who is unmarried, and Janet, who married relatively late in life, it's an ambiguous story: It reflected their mother's ideas about a woman's place, but also her desire to keep the twins humble—surely, she knew they were listening.

Bea Witzleben, an assistant US attorney in Philadelphia and one of Mary's closest friends, always hears something else in the story when the twins tell it. She hears the sudden hush, the immediate silence that betrayed how desperate the twins were to hear their mother's answer. "At that point, they're two accomplished women—but they are still so desperate for their mother's approval," she said.

LEADERSHIP AT THE LEVEL ACHIEVED by the Murguías entails living a public life. But public life also means public scrutiny. And when public figures stumble, err, or blunder into disgrace, their families—even resourceful, problem-solving, supportive families like the Murguías—do not have the option of managing those moments privately.

"A federal judge in Kansas was given a rare public reprimand on Monday for sexually harassing female judiciary employees and having an affair with a felon that made him 'susceptible to extortion,'" reported an article in *The New York Times* on September 30, 2019.

The subject was Carlos Murguía. His siblings were more or less blindsided. They had known for a long time that their brother's personal life was under strain, but they had been unaware of his outright infidelity (with a felon, no less), and knew nothing of his unprofessional conduct until he called to tell them shortly before the issue became public. In the following weeks, Carlos was uncommunicative about the details of the situation even with Ramón, to whom he spoke almost daily, two lawyers treading carefully around sensitive material.

The details turned out to be grim. Carlos had sent sexually suggestive texts to court employees, the story reported, even after they had asked him to stop. At the time of the reprimand, Carlos apologized and agreed to take "certain corrective actions," which resolved the issue at the time; but another investigation followed, with subsequent calls in the House Judiciary Committee for greater safeguards against sexual misconduct on the bench. In February 2020, Carlos, under threat of impeachment, resigned, his good reputation erased, his income and pension gone, his marriage soon finished. He moved out of the house he'd lived in next door to Ramón.

Carlos, whom I met briefly in person on an early visit to Kansas City, responded only in writing with carefully worded responses when I approached him to talk after the news broke. "I failed to respect workplace boundaries between employees and supervisors," he wrote. "I apologized to all the victims of my regrettable actions and conduct. I am truly sorry."

But the events seemed to crack something open in the siblings. Each felt a sense of bewilderment and confusion, trying to understand how they knew so little about someone so close to them. Had Carlos changed over the years, and, if so, why? Or had he been someone they never really knew all along? They didn't doubt the allegations, but they couldn't fit them into any conception of who he was. "I want to explain my brother Carlos to you in some way that will make sense to you," Ramón told me. But he could not.

Even more notable, to me, was the siblings' sense of devastation—and responsibility—as a family. "It flips you back on your heels," Janet said. "*This happened to Carlos—so why am I feeling this way?*" she asked herself. "*What's going on here?* We're individuals—but we have always identified as a cohort." They had all benefited from the story of their unusual family success, a story that now felt compromised. "It was part of our narrative, part of our story—so how do you explain it when . . . something else happens?" said Janet. "One person's aberrant behavior tears into that—what does it mean?" Ramón shared the same feeling of injury. "We all had the breath knocked out of us," he said. "We have all been hurt."

"I owe you an apology," Mary told me when we first spoke after the news broke: She worried that she had held out the Murguías as some kind of ideal, in error or in pride. "We are proud of our accomplishments, of being a credit to my parents," she said. "When people are pointing to us as examples, you want to convey, 'You can do it.' But you do it all with a qualification that we are not perfect. We got into a groove—getting recognition, you kind of like it." Now she felt guilt, even shame. "Even though it wasn't my conduct, it was devastating. I felt like: *Did I contribute?*" she said. "I know I didn't. So then why do I feel so terrible?"

The family narrative adjusted to accommodate a simple truth: "We're flawed as human beings, no matter how much we want to adhere to those higher standards," said Ramón.

Even as Carlos was paying for his abuse of power, Mary was reinforcing laws against sexual harassment by reversing a lower court's opinion that Ashley Judd had no grounds to sue Harvey Weinstein for sexual harassment. Weinstein argued that Judd—toward whom he'd made aggressive sexual advances in a hotel room where she'd expected to have a business meeting—was not formally his employee. Given Weinstein's power at the time, Mary ruled in 2020, "The relationship was one that would have been difficult to terminate 'without tangible hardship' to Judd, whose livelihood as an actor depended on being cast for roles."

For most of their lives, the Murguías shared a common faith in their own competence, which, combined with their intelligence, diligence, and agility, had served them well. But perhaps in Carlos's case, this confidence had given way to something more perilous—a sense of invincibility. "Ego," said Mary, who, of the other three younger siblings, perhaps suffered the most from the news, as a fellow judge; at the time that Carlos resigned, she was serving—now, with some self-consciousness—on the Judicial Code of Conduct Committee, an ethics committee for the federal judiciary. "I know he thought, of all of us, he was by far the smartest, the superior one. But I thought he had enough humility that it wouldn't be a problem," she said. In time, she'd come to think the problem was less easily explained—that her brother had profoundly "lost his way," as she put it.

The hint of resentment in her initial response—that Carlos always harbored the belief that he was the smartest of the youngest four—once again invoked an old familiar dynamic, the constant reckoning among the siblings about where they stood in relationship to one another. Carlos, in the email he sent me, said he was trying to protect his siblings by shielding them from the details of the investigation, which was no doubt true. But he added that he was also "embarrassed to tell them about my disappointing conduct."

Carlos's actions filtered into his siblings' professional lives, where they fielded awkward questions with carefully worded answers. Janet had staked her own reputation on advocating for his appointment to the bench; that was on everyone's mind, even as the four siblings, including Carlos, continued their regular late-night calls, siblings trying to make up for each other's blind spots or gaps in knowledge or contacts.

Always the most reactive, Mary especially felt outrage, and a need for some kind of reckoning. "What in the world were you thinking?" Mary asked Carlos on one of those calls. At some point, Carlos said he was trying to stay strong for the sake of his children. "Now you're thinking about your kids?" Mary asked, before Ramón and Janet interrupted her, trying to de-escalate.

It was always natural for the Murguías to credit their success to their upbringing; as to how to explain their failings, the answer would be less straightforward. An emphasis on achievement meant that when reckoning with human and fallible impulses, the shock was genuine; in general, they had a hard time talking about feelings and frailties. "We were connected," Mary said, "but not in a way that shared a lot of emotional feelings."

In Mary's role as a prosecutor, she sometimes worked on cases involving child abuse, which brought her into contact with child psychologists. The contact was as close to therapy as any of the siblings had ever come, as far as she knows. She was struck by a technique that psychologists used with young people to help them identify their feelings when they were having trouble expressing them. The tool was simply a piece of paper that showed various faces, each of which expressed a different emotion—happy, angry, worried, sad, confused.

Mary looked at the paper as if it were a revelation. "Could I have one of those?" she said to the psychologist. She was just self-aware enough to know that she had trouble analyzing her own turmoil, her own emotions, and even more describing her feelings out loud.

The Murguía upbringing had been so intense, so competitive; it had been loving and suffocating, filled with teasing and laughter, a continual ruckus that might have taken the place of open communication, the kind that can foster self-awareness. Mary did not excuse any of it; but she tried to understand it. That undercurrent of competition among the siblings, in everything from board games to academics, had given her valuable skills she took into her professional life—but it also came with a cost. "When people get from one point to another," said Mary, "if they achieve any success, usually there is some sacrifice that causes ramifications that are difficult. Some loss." That loss could be in self-reflection; it could be in perspective; it could be in peace of mind; it could be in love.

Amalia and Alfredo both died long before Carlos resigned. At Alfredo's deathbed, with the siblings gathered around, Janet had leaned in to say what all of them felt: "Dad, I love you," she said. He was silent. Janet leaned in again. "Dad, I said I love you," she said. Her father finally spoke: "I like you, too." Mary and Janet, when they relay that story, sometimes laugh until they cry.

The siblings tried to reconcile their outrage with the softer virtues they learned from their mother. "We're lawyers—so that part of us is calling for accountability, accountability, accountability," said Janet. "But the other side of my brain is telling me what my mother taught me, which is that we stick together in good times and bad. My mother was all about forgiveness, about giving people a second chance."

When Mary was still living at home, she once asked her mother to pray for her, because she had a big closing statement to deliver the next day in the prosecution of a violent sex criminal. "I'll pray," her mother told her, "but for the criminal."

On one of their regular family calls, Ramón suggested that Carlos apologize not just to the public and to his family but also personally, to some of the neighbors with whom they had all grown up. Carlos told him he'd already resolved to do so. In the following months, Carlos went door to door, apologizing to neighbors, many of whom knew his parents; he wanted them to know that he was aware that his parents had raised him to be better than how he had behaved.

Mary, who became chief judge of the Ninth Circuit in 2021, announced that addressing sexual harassment in the judiciary was one of her chief areas of focus, leaning into the issue, rather than shying away from it, all the while knowing that her own brother's case might turn up as precedent. In 2024, investigating a tip, Murguía oversaw the investigation of the Alaska federal judge Joshua M. Kindred for sexual misconduct, ultimately filing an order that led to his resignation.

Talking to Carlos's siblings, one does not have the sense that they have excised him from their story to preserve their own sense of excellence. If anything, his experience deepened their self-reflection about their own lives. And they continue to support Carlos, as their mother would have wanted. They brainstormed about what kind of work he could do to remain on his feet and remained in close contact. He watches the Kansas City Chiefs game with the family when they are in town, and he celebrates holidays and birthdays with Martha, Rose Mary, Alfred, Ramón, Janet, and Mary when all of them gather with their respective families at the home on Twenty-seventh Street. It is located just catty-corner from the house in which Carlos grew up so humbly—bigger, more modern, but on the same block. It was still the family house, and he was still a Murguía, which meant he could always go home.

SIX

Luck and Fate

In the book *Success and Luck,* the economist Robert Frank makes the case that successful people often mistakenly downplay the role of luck in their lives, underestimating its role while overvaluing their talent. That analysis, he further argues, often yields a sense of entitlement that can, multiplied across entire classes of people, have profound and troubling effects for social policy. His book is a thoughtful meditation on the lucky break—a factor that is recognized and embraced, in other words, by the people who are not just successful but wise.

For the Holifields, that lucky break came in the form of a letter, one that expressed hate, but nonetheless landed them a new assignment, in Tallahassee. For the Brontës, also, a troubling development may have provided a lucky break that set so much else in motion.

The concern for the Brontë family was Patrick's eyesight, which was failing fast by the time he'd reached his mid-sixties. A surgeon in the family's circle assured them that an operation could spare some of his vision. Charlotte traveled with him to Manchester for the procedure late in the summer of 1846. During his recuperation, he was prescribed silence and darkness. He was not to talk, nor to be read to.

Charlotte's letters make it clear that she was concerned for her father, and yet the month or so she and her father spent in Manchester nonetheless offered her a rare luxury: time and quiet to think and write without the tug of other obligations. She was not being interrupted mid-thought, as so often occurred when she was working as a governess. (In one journal entry, written when she was working as a teacher, she complained of a "dolt" who interrupted her in the middle of a creative reverie, a distraction so annoying, she wrote, "I thought I should have vomited.") In Manchester, she could be both dutiful to her father and have the space to indulge her imagination.

That moment was opportune: Charlotte was poised to do her best work. She had already spent much of her childhood writing competitively and collaborating with her siblings, especially her brother, Branwell, creating those imaginary worlds. As a young woman, she had already completed a first, lesser novel, *The Professor*. That book had not yet found a publisher, but she had received just enough encouragement to take her own talent seriously. She was likely inspired by her sister Anne's novel *Agnes Grey*, which had not yet been published, but which gave dignity to the voice of the lowly governess; perhaps she thought she could even elevate the subject further. In Manchester, she started writing *Jane Eyre*.

"Sitting by her blinded, silenced father, she dares to take up her pencil and write for the first time in her own voice," Sheila Kohler writes in *Becoming Jane Eyre*, a novel that tries to understand how Charlotte's time in Manchester proved so creatively fertile.

Great writing can happen in stolen moments. But for Charlotte, her father's mandated silent retreat may have offered her the emotional space she needed to begin to explore the personal losses and injustices she drew upon for *Jane Eyre*. Relying on a first-person voice, she dipped in and out of the most salient experiences of her life—including the loss of her oldest, beloved sister, Maria—to tell an irresistible story, in a novel that spoke of the passion and power of resistance that could be found even in plain women of humble means.

When *Jane Eyre* was published in the fall of 1847, just over a year after she started writing it, the novel became a runaway success—its reader felt "soul speaking to soul," as one prominent critic put it in a glowing review. Its success spurred the publication of Emily's *Wuthering Heights* and Anne's *Agnes Grey,* as their publisher rushed to capitalize on two more authors who shared the same pseudonymous last name (all three sisters published under the last name Bell). That unusual respite in Manchester provided the opportunity from which so much literature and culture would follow.

For the Murguía siblings, the lucky break is an explicit part of their shared history, not equal in weight to the power of faith, family, and community, but a phenomenon that demands recognition nonetheless. The coin toss is the stuff of Murguía family legend, a story often told for laughs, sometimes with dark undertones. The role of chance ultimately speaks to the nature of siblinghood—how sisters so closely bound could be raised in the same home and yet see their life trajectories diverge. The coin toss is the outside world asserting the power of randomness on even the most strong family culture.

NANCY SEGAL, A TWIN, and a professor of developmental psychology at California State University, Fullerton, conducts research in the field of behavioral genetics. She has much less interest in luck than she does in the interplay between genes and the environment in influencing human behavior—but chance plays some part in those differences, too. And when I came to know Nancy some years ago through a case study that interested both of us, luck seemed to be a prominent theme that could not be ignored. It also appeared to have played a prominent role in her own life as a twin, as she told it.

Nancy, who grew up in Philadelphia, vividly remembers heading to the first day of kindergarten with her fraternal twin sister beside her. Nancy watched as her sister was brought to a classroom run by a young, beautiful teacher. Nancy was then led to her own classroom, where the teacher, she felt, was less appealing; she can remember the brown color of her teacher's stiff, dowdy dress. She recalled sobbing inconsolably until, finally, another teacher brought her sister over to comfort her. But the two remained in separate classrooms for the remainder of that year.

The girls' school experience was different in the years that followed, as well. Nancy had so much energy as a young girl that her teachers sometimes bridled. She did not want to nap; she did not want to eat at snack time. What she wanted to do was talk. Her sister's teachers did not face the same challenge. In third grade, the girls were separated when the school divided the classrooms by aptitude. Both girls tested highly enough to enter the most competitive of the classes, but the school opted to put Nancy's sister in the first class and Nancy in the second, rather than have them competing in the same class, or conspiring, or even fighting. (Nancy Mitford, one of the six historically significant Mitford sisters, once said, "Sisters are a shield against life's cruel adversity," to which her sister Decca responded, "But sisters *are* life's cruel adversity!")

Nancy said that she did not mind the decision at the time, but, in retrospect, she has come to find it wrong. "It was an injustice," she told me decades later. "I think it was a severe and serious mistake. They should have found another school for me." Not that it mattered in the long run: She would find her own places to shine, in academia, on television, in books, in interviews. But she still feels the sting of injustice and some pang of loss when she reflects on it. How a child feels in the moment and when she reflects on it later in life—both are meaningful in their own right.

In sixth grade, by which point their family had moved to New York, both girls tested for entrance into one of New York City's public, citywide schools for gifted and talented children. Whereas all the students in her sister's class were instructed to take the test, Nancy was one of only three in her own class who were chosen to do so. That might have been the opportunity for Nancy to establish herself as her sister's academic equal, or at least an academic standout in her own right; but early on, the day of the exam, sparks nearly flew. In a chemistry classroom, Nancy accidentally leaned on a lever attached to a gas jet, and fumes unexpectedly came pouring out. Hearts raced, senses recoiled, and many, possibly all of the students (Nancy's memory fails her on this point) were evacuated out to the street. Her sister was also present in the classroom that day, perhaps most notable for being the Segal twin who did not wreak havoc on test day. It was one of those chance moments that had consequences: When everyone finally sat down to take the test later that day, Nancy, still rattled, did not perform well enough for entrance to the school for which they were testing, Hunter College High School. Her sister did.

Nancy went on to have a perfectly happy academic experience at her local junior high school, and then tested into the Bronx High School of Science. Ultimately, her sister, who graduated at the top of her class, went off to an Ivy League school, fulfilling a dream of Nancy's father, who had longed to go to one himself but instead attended Boston University, where Nancy enrolled.

The discrepancies of fraternal twins lay bare their parents' inability to render the world perfectly fair, no matter how much love, nurturing, and extra support they give. Seen from a wider perspective, the Segal twins' fortunes were essentially the same: They both came from a middle-class, well-educated family and had the additional advantage of attending prominent universities, already rendering them more similar to each other than to the vast majority of the population. Both went on to have fulfilling, successful careers. Yet within the cloistered world of competitive public high schools in New York, an entrance to an Ivy League school was tantamount to an anointing. The twins' parents may have been sensitive to some of these kinds of differences: When Nancy's sister was admitted to Hunter, no family celebration followed. They waited for Nancy's admission to Bronx Science to reward both girls with a special dinner.

In the end, Nancy said, Boston University was more than good enough, especially when she found her calling senior year: studying twins. She had taken a class in abnormal psychology that required a paper about personal development, and chose to write about the separation of twins in classrooms, clearly inspired by her own experience, which she characterizes as formative. Was the trauma in the separation or in the feeling of unfairness in teacher assignments? Just as Mary Murguía was penalized for being a twin, Nancy lost out on being in the most challenging classroom because of a policy that separated twins. (Nancy has come to oppose mandatory policies in schools that enforce the separation of twins and believes the decision should be made on a case-by-case basis.)

Having found a topic that fascinated her—twins—Nancy finished college with strong grades, and that spring she was admitted to the graduate program in social sciences at the University of Chicago. (She was so overjoyed that she threw the letter in the air, then spent fifteen panicked minutes unable to find it, half-convinced, briefly, that no such letter had ever even arrived.)

By the early 1980s, Nancy had landed as a postdoctoral fellow at the University of Minnesota, where the psychologist Thomas Bouchard, Jr., was studying identical twins who had been separated at birth and then reunited in adulthood. In systematically comparing their traits and talents, he often found that, regardless of how different the twins' home environment had been, they predictably showed remarkable similarities beyond their physical appearance. In the years since, Nancy has become one of the people responsible for popularizing that academic research, a body of work with methodology that is still contested and critiqued, but that has also helped normalize the acknowledgment of some genetic influence in the development of personality traits and behavior. One of Nancy's books, *Deliberately Divided,* features the story of the triplets who were the basis of the documentary *Three Identical Strangers,* about three identical triplets separated and raised in different homes for the purposes of research; another, *Born Together—Reared Apart,* tells the story of, among others, the Jim twins, who'd been separated at birth but grew up to be adults who vacationed in the same three-block ocean town, who smoked the same cigarettes, who married women with the same first name, who shared, in other words, an eerie overlap of quirks and habits that seemed almost preordained.

Nancy has devoted her life to research—more controversial when she started—that argues for a role of one's genes in one's fate. But her own story is replete with injustices or quirks of fate that she sees as significant, if unquantifiable. The ripple effects of those chance decisions and happenstance occurrences—the assignment of kindergarten teachers, the gas shooting into her chemistry classroom—are not easy to capture in scientific metrics but could nonetheless be consequential, intersecting with other unpredictable factors, creating subtle results that no algorithm could capture.

Because of her own upbringing, Nancy was also interested in studying the dynamics of twins themselves. She and her sister did not fight, nor did they have an unusual closeness; their interests diverged, as did their friends. "Sometimes I complained to my mother, 'I have no one to play with,' and my sister would be sitting right there in the next room," recalled Nancy. Identical twins, on the other hand, seemed to her intimately connected, enviably so—she could see it in the amount of time they spent together, even in their body language.

In 1993, Nancy published research that found that grief, in fraternal twins who lost a co-twin, eased more quickly than it did in identical twins. Since then, research has shown that identical male twins live longer than fraternal male twins, for reasons that the researchers theorized might be explained by the protective effect of having a life partner more closely looking out for one's well-being.

Nancy's objective was to see whether identical twins could be captured, in the moment, behaving in some way that was different from the interactions of fraternal twins. Her study design involved a puzzle, which depicted a dog beside a bone and was suitable for the children in the study, the youngest of whom were about six. The challenge for the various pairs, both identical and fraternal, was to complete the puzzle together as quickly as possible. As the children were filmed, their faces were obscured on the videos that the evaluators watched, so that they could evaluate the quality of the interaction—how close the twins' bodies were, whether they struggled over positioning—without knowing if they were observing identical or fraternal twins.

There may well have been other clues—discrepancies in height or weight—that could suggest which kind of twin pair was being observed (Nancy said any differences were minimized by the children's being seated at a table). But the study found that the identical twins completed the puzzle more quickly than the fraternal twins did; and the evaluators often described the identical twins as working in concert, as if taking part in some kind of gentle, coordinated dance. The fraternal twins had more struggles over the toy and displayed fewer signs of amiability.

One might think that identical twins would be more likely, not less likely, to compete and fight for attention, dominion, or individuation. But in this study, and others that explored similar themes, Nancy found more cooperation and detected more altruism between identical twins than she found in fraternal twins or siblings.

At times, the relationship between identical twins can seem so close, so giving, that it surpasses even the most generous impulses of typical loving siblings. In 2014, Tracy Barnes won a spot to compete at the Olympics in Sochi in the biathlon (a sport that combines shooting and cross-country skiing)—then gave it up so that her identical twin sister, Lanny, who had been sick during selection races, could compete in her stead. "I think of it as just transferring it," Tracy told NPR at the time. "I'm still in a way going to Sochi, it's just I'm going through her." And Lanny knew her twin well enough to believe her. "I know Tracy better than anybody," she said, "and I know this is what she wants. I just want to make Tracy happy."

In identical twins, one sees the dynamics of two humans who would be one, were it not for the entirely random splitting of one embryo into two, a function of pure chance.

EVER SINCE I BECAME the mother of fraternal twins, I have had a keen interest in the subject of all twins, which is only one reason I was drawn to the fascinating story of two pairs of twins from Bogotá whom Nancy started studying in 2014. It was a story of mistaken identity, an adoption story, a comedy, a tragedy. And it was a parable that shed light, even for Nancy, on how to think about how one's environment affects one's chances at success.

The remarkable story started in late December 1988, when one set of identical twins was born in Bogotá and another set of identical twins was born in the remote rural region of Colombia known as Santander. One of those Santander twins—Carlos—was so ill after birth that his aunt brought him to the hospital in Bogotá. There, someone must have laid Carlos alongside one of the newborn identical twins from Bogotá by accident. Instead of going home to Santander, Carlos would remain with that twin in Bogotá for the duration of his childhood; in Carlos's stead, as part of this mix-up, one of the identical twins born in Bogotá, William, was taken to Santander.

Now, the twin pairs were scrambled: Carlos, the twin who had been ill, and who came from Santander, would be raised in Bogotá with a Bogotá-born twin named Jorge. And William, who should have been raised in Bogotá alongside Jorge, would instead be raised in Santander alongside Carlos's identical twin—a child named Wilber.

Carlos and Jorge, growing up in Bogotá together, were so different from the outset that people assumed they were fraternal twins; the same was true for the mismatched brothers growing up in Santander, William and Wilber.

In their mid-twenties, the young men raised in Santander (William and Wilber) made a fateful move: They came to Bogotá to work at a butcher shop. By chance, someone who worked with Jorge, one of the twins raised in Bogotá, happened to visit that butcher shop, where she saw someone who looked so much like her colleague that she couldn't shake the similarity. The connections unfurled from there, and all four young men eventually met. All four were shocked to learn that the brother they thought was a fraternal twin in fact was not a blood relative at all. And two were forced to confront the pain of knowing that their mother was not their mother by blood—that, by rights, they should have grown up in another family far away. Chance had intervened, dramatically altering the trajectories of all four lives—and, for Nancy, casting into relief the competing effects of nature and nurture, genetics and environment.

The upbringing of the two sets of twins was about as dramatically different as can be found in the literature on adopted or separated identical twins. The two boys in Bogotá went to strong schools, and enjoyed exposure to books, magazines, televisions, computers, and video games. The boys in Santander grew up in a remote, rural area, working for hours in the sugarcane fields, living in a home with limited plumbing and little to no academic stimulation. After years of making a hilly trek to school, often through deep mud, they stopped attending at age twelve. Their new high school was simply too far away. College was not an option. They both endured the hard labor of military service in Colombia.

Despite those differences in upbringing, the young men were strikingly similar to the identical twin from whom they'd been separated.

Jorge and William, the separated twins whose biological mother was from Bogotá, shared personalities that their friends invariably described the same way: They were both sociable, sensitive, and affectionate, exceptionally charming, even if they were both desperately awkward dancers and perpetually late. Wilber and Carlos—whose biological mother was from Santander—were both punctual, more brusque and moody, but still appealing enough that acquaintances felt compelled to win them over; both were noticeably vain about clothing and were superb dancers.

But the experiences of the separated identical twins also diverged in dramatic ways that laid bare how environmental effects can swamp whatever genetic predispositions a person might have. Nancy, who administered a battery of tests to both sets of twins, had expected that the intelligence scores of the identical twins would be similar if not identical to each other; in fact, especially for Jorge and William, that was not true—their results were closer to those of the siblings they grew up with than to each other. William, who stopped attending school at age twelve, had little patience for the aptitude tests. It was not only the answers, but the concepts at the heart of the questions, that eluded him. He lost motivation and energy to continue the tests; he was deeply uninterested, even baffled. His results, not surprisingly, were much lower than those of his identical twin, Jorge. Most likely, the tests could not capture whatever intelligence was obvious in William, the talent that would eventually help him, years later, accomplish his objective to graduate from college with a degree in law; it seemed to measure mostly his success in taking a test the likes of which he'd never seen before.

William's experience reflects one of the enduring critiques of IQ tests, which is that they reflect class and opportunity rather than some innate intelligence. Julia Leonard, of the Leonard Learning Lab at Yale, often illustrates the problem by referencing one of the questions on an IQ test she regularly gave as a student. The test directed her to ask students, "If you find a wallet at a store, what do you do?" The answer, she said, was return it to the store clerk. "And a lot of low SES [socioeconomic status], Black kids were like, 'Don't touch it—leave,'" she said. Those young people, who had rational reasons to think they might not be trusted, were docked points.

Perhaps more profoundly, in outlook on life, William and Jorge were fundamentally different, suggesting that, in that regard, their environment actually had shaped deeply held perspectives. "Nothing in life is easy," William once told me, a sentiment that captured the grueling labor his upbringing entailed, the hours of sweat and toil in the fields where he farmed, the physical exhaustion he would have felt just from walking, alone, as a small child, for hours to and from the closest store. It was impossible to imagine his identical twin, Jorge—sunny, optimistic, a cheerful and confident young man who grew up in Bogotá—expressing the same thought.

When I met those four young men, they were grappling with a barrage of revelations that challenged everything they thought they knew. Simply confronting a genetic replica of oneself would be disorienting, but they were also being forced to confront everything they had lost. William was the much-beloved main emotional support of his (unwitting) adoptive mother in Santander; but he never got to know his biological mother, who died before the two sets of twins were united. He grieved that loss; he also grieved the time lost with Jorge, his identical twin, with whom he did feel a powerful and loving connection almost immediately. He had longed to go to college and grieved the relatively easy access he would have had if he had been raised in Bogotá, his rightful home.

I sensed that the young men all felt some anger—the feeling that their lives were the culmination of some massive cosmic joke. To be a twin is to feel more keenly than most the power of random twists of fate: It is true for fraternal twins (Why does one stutter but the other doesn't?); and it is true for identical twins (Why did my sister win the coin toss, and I didn't?). Staring at his identical twin for the first time, Jorge, a trained engineer, told me, "was like looking into a mirror, and on the other side of the mirror was a parallel universe." For Carlos, seeing his identical twin, undereducated and working in a butcher shop, forced him to confront the possibility that he, too, might have ended up with little schooling—that had he been raised in Santander, he likely would have been working behind a butcher counter or, worse, might have been lured to fight with leftist guerrillas, as so many young men from Santander were. He might have ended up dead in one of the many skirmishes common in the area at the time.

Confronted with his less advantaged identical twin, Carlos, an upwardly mobile accountant, had no choice but to see how much the relative privileges of his upbringing—his schooling, his leisure time, even his access to a speech therapist who cured him of a speech problem that lingered in Wilber—contributed to his success. That so much of his luck would have been forever obscured, but for the chance encounter of a friend of Jorge's with William at the butcher shop, might have been the most unnerving twist of all. So much of what influences people's chances for success in life is not just discounted but completely hidden from sight.

THERE ARE FORCES—countless coin tosses beyond a family's control—that influence how various siblings within a family fare. Even within the family, there are likely powerful dynamics creating differences that are not within the control of the parents.

For many years, the influence of birth order has been a seemingly obvious factor onto which most armchair sociologists could easily map some kind of logic. The subject of siblings' birth order has lured researchers since Francis Galton was first tracking eminent people and noticing that, according to his research, a majority of them were firstborns.

Sigmund Freud, an oldest child, believed that "the position of the child in the family order is a factor of extreme importance in determining the shape of his later life and should deserve consideration in every case history." But, beyond that, he did not dwell on birth order, considering sibling relationships far less influential than the parent-child dynamic. His intellectual successor, fellow Austrian Alfred Adler—the younger sibling of a brother who happened to be named Sigmund—established, among other things, a fully formed theory of birth order that sought to explain why siblings raised in the same family were often so different. "It is a common fallacy to imagine that children of the same family are formed in the same environment," he wrote. "Of course there is much which is the same for all in the same home, but the psychic situation of each child is individual and differs from that of others, because of the order of their succession." The younger sibling of an older brother who, in their youth, surpassed him in academics and athletics, Adler asserted that middle children were the most psychologically healthy, having been forced to share resources and attention. (Younger siblings, he also believed, who learn from observing the mistakes and successes of their older siblings, not just from their own experience, make for the best therapists.)

In 1996, in an era when classical psychoanalysis was starting to lose cultural ground to psychiatric medicine and behavioral therapies, the topic of birth order seemed to offer a more accessible psychological explanation of the self, rising up to dominate popular psychology through a 1996 book by Frank Sulloway, *Born to Rebel: Birth Order, Family Dynamics, and Creative Lives*. The book, a discursive and erudite attempt to explain how one's place in the family shapes one's identity, was a runaway bestseller, praised by intellectuals such as evolutionary biologist Edward O. Wilson, who called it "one of the most authoritative and important treatises in the history of the social sciences."

And yet the book's findings—that the oldest children, essentially, embraced traditional forms of authority, while the youngest, seeking a niche of their own, were rebels who made way for change—did not put to rest more than a century of debate on the influence of birth order. (There is even a highly debated body of literature on the birth order of scholars of birth order.) "Rebel Without a Cause or Effect" was the name of one prominent paper that challenged Sulloway's thesis in 1999. The article, cowritten by sociologist Jeremy Freese, now of Stanford University, and published in the *American Sociological Review*, took issue, in particular, with Sulloway's assertion that birth order was a better predictor of a person's social attitudes than was class or gender. Freese critiqued Sulloway's methodology (which relied largely on historical data about Western white men), but he also analyzed the results of a survey of social attitudes—toward race and various political issues—of some two thousand respondents in a survey conducted by the National Opinion Research Center. Race, age, gender, level of parents' education—all of those, Freese found, were better predictors of social attitudes than where someone fell in the birth order. Sulloway had argued that most revolutionaries were youngest children. But Freese found that one's place in the socioeconomic hierarchy, not in one's individual family, was what mattered most when it came to how conservative or progressive one was. Just as behavioral genetics was starting to challenge the primacy of psychosocial family dynamics in forming a person's personality, sociologists were making the case that structural forces—economic inequality, racism—had more power to shape attitudes than how siblings and their parents interacted in the home.

Major studies in 2015 and 2019 also failed to identify a significant impact of birth order on personality traits like conscientiousness or risk-taking. In both cases, researchers wielded huge datasets that were unavailable to their predecessors, some of whom had relied entirely on the opinion of one sibling to rate the personalities of others in his or her family, or only asked interview subjects to compare family members' personality traits to those of other siblings in the same family. The flaw was apparent: The oldest child might be conscientious relative to his younger sister, for example, but not particularly so compared to the population at large. Brent Roberts, one of the scholars behind well-designed research that challenged findings of birth-order effects on personality, points out that "within-family" studies have an essential flaw: "The oldest child is always older." The older child is more responsible than the younger sibling, it's true— but that's an underappreciated function of age rather than innate personality. The fact that a thirty-two-year-old thinks to call home more often than her twenty-seven-year-old younger sister does not mean she's especially conscientious relative to everyone else (or especially conscientious in ways that translate to other expected outcomes, such as how much education she acquires or how wealthy she'll become). In papers that corrected for these age-related disparities or relied on between-family comparisons, the birth order effect on personality all but disappeared.

And yet researchers continue hunting for the power of birth order, including the Columbia University economist Sandra Black, who was able to access Swedish military records that included the results of screenings for cognitive skills as well as for a given conscript's "capacity to fulfill the requirements of military duty and armed combat," as Black wrote in her paper "Born to Lead? The Effect of Birth Order on Non-cognitive Abilities," which was published in *The Review of Economics and Statistics* in 2018. The trained evaluators for the Swedish military were asked to rate conscripts on a variety of measures. Was the person before them likely to be emotionally stable, persistent, socially outgoing, willing to assume responsibility, able to take initiative? Black analyzed a sample of more than half a million men. The first-borns scored higher on cognitive tests but also scored higher on the general rating used to designate a conscript's overall likelihood of succeeding in the military. "We used that measure to try to get at personality, and what we find is that, again, it is better to be the firstborn in the family," said Black in an interview with *For All,* a Federal Reserve–affiliated magazine, about her work.

What traits constitute "better" is, of course, philosophically debatable—and some researchers argue that the effects Black observed are so small as to reinforce birth order's irrelevancy when it comes to shaping personality. But Black also found evidence of beneficial outcomes (a phenomenon distinct from personality) within her cohort of older siblings, including greater longevity and higher incomes. Analyzing a database of professional data, also from Sweden, she found that firstborns were 30 percent more likely than third-born siblings to rise to the top ranks in corporate environments. "They're more likely to be in positions requiring leadership ability—managers and the top managers in the company," she said. "They are also more emotionally stable and willing to assume responsibility than later-born children. And we find that it's systematically declining with the birth order." Another paper, also based on Swedish data, found, in 2019, that older siblings were significantly more likely to run for office and to win—and that birth order was a better predictor of that success than gender or even education level.

It seems plausible that great ambitions take root earlier with the oldest siblings, quite possibly because parents start dreaming big for the next generation as soon as the first one is born. Such was the case for Joseph P. Kennedy, Jr., John F. Kennedy's older brother, of whom great things were expected from the start (he died while serving in World War II). "Well, of course, he is going to be president of these United States," Joseph Kennedy, Jr.'s grandfather, John Francis Fitzgerald, predicted to a reporter for *The Boston Post* in 1915. "His mother and father have already decided."

There may be rational reasons for parents to exert more pressure on their oldest children, concluded, in 2015, Joseph Hotz, an economist then at Duke, who found measurable ways in which parents did treat their firstborn children differently. Parents exert the most pressure on those children to succeed academically, enforcing more rules and punishments in the event of failure, Hotz's research showed. An analysis of data that tracked twelve thousand young individuals and that involved two rounds of interviews with their mothers found that the oldest child was more likely to have homework monitored and more likely to have limits placed on typical pleasures that compete with homework time, such as watching TV. The effect was proportional: The more children there were in a family, the more pressure the parents put on a first child. In Hotz's analysis, the investment in disciplining the oldest child (the stress of the conflict, the effort it takes to monitor that child or to mete out and enforce a punishment for slacking) is well worth it, if those behaviors establish a "threat"—the reality of consequences—and encourage the oldest to set a standard and an example. Economists seek to identify rational responses to costs and benefits: In Hotz's analysis, it would be logical for parents to go easier on the younger siblings, not because they're favored or babied but because, with fewer children left for them to influence in turn, the hassle of reinforcement reaps fewer dividends. He also found that the pressure put on oldest children seemed to be effective. His work confirmed that the oldest children in the families he studied were the academically strongest among their siblings. (Rose Kennedy, matriarch to John F. Kennedy and his eight siblings, seemed to have intuited some version of this on her own. "If you bring up the eldest son right, the way you want the others to go," she wrote in her memoir, *Times to Remember,* "that is very important because the younger ones watch him. If he works at his studies and his sports until he is praised, the others will follow his example.")

In his book *The Pecking Order,* the sociologist Dalton Conley was skeptical about any theories of birth order related to parenting or family role—he argued that economic and situational factors at any given point in time within a family were far more influential. To the extent that birth order matters, he contended, it is usually because of the way resources are distributed, rather than because of any internal psychosocial dynamic of the family. In families with three or more children, middle children, for example, are less likely than the oldest or the youngest to attend college. Conley theorized that this could be because they have neither the benefit of the first bite at the financial apple nor the benefit of parents who are at their oldest point in parenting, which is when they are often at a point of greatest financial stability.

One of the reasons that studies of birth order, and siblings in general, tend to frustrate researchers is that family structures are messy, with families combining and confounding basic notions of who's even first: Siblings are split between divorcing parents or experience other forms of trauma or upheaval at different moments, some less crucial than others. In the Murguía family, Martha was the oldest; but because of her disability, Alfred became the de facto oldest child. And yet, even there, the role of birth order in the Murguía family may have been confounded by more powerful countervailing factors: The earliest years of their lives coincided with the years of greatest poverty and isolation for the Murguías, a time when Alfredo was working as a day laborer. Alfred, Rose Mary, and Martha all spent the first years of their lives in much more challenging circumstances than their younger siblings, in that home where they struggled even to keep warm.

The oldest Groff, who became a serial CEO, was considered by Lauren to be the most academic of the three siblings, although Sarah considered Lauren a close second if not Adams's equal (so much so that Sarah, as she herself theorized, while always a strong student, pivoted to the pursuit of athletic glory). Bishop Holifield, too, who was general counsel for FAMU, technically held a more classic leadership position than either of his siblings. But could it be that Bishop—whose parents considered him the most brilliant, according to Marilyn, and who was valedictorian at FAMU—set the standard for his siblings? Possibly. If early development matters, would it make sense to think of him as the oldest child when, in fact, his life overlapped, all too briefly, with an older sister?

Charlotte Brontë functioned as the oldest of the Brontë family—she was practically the business manager of the Brontë writing enterprise at its outset—but she, too, survived two older sisters whose influences remained with her for the duration of her life. And she held little sway over her younger brother, the fourth child born into the family, whose role as the only boy was surely more relevant to the family dynamic than his birth order.

SUPPOSE THAT IT'S TRUE THAT parents hold oldest children to a higher standard, knowingly or not, in the hopes that the oldest child will then set a high bar for the younger ones to meet. That theory makes sense from an efficiency standpoint but puts a high burden on parents hoping to see all their children thrive academically: It's up to them to figure out how best to elicit the oldest child's high performance, which is known to be a complicated, elusive, and potentially fraught project.

Researchers have found one way of improving school performance that seems more reliable than so many that rely on execution or have very short-term effects or sporadic results. It does not demand the effort of daily decision-making—how much to punish, how much to enforce—but instead involves one simple fact, which is the age at which a child starts kindergarten. That timing, it turns out, is not meaningful just for that child—it can also be meaningful for that child's siblings. Many parents see the age at which their child starts school as a matter of chance: When did they happen to be born? But others see how they manage that variable as a matter on which they should bring their own agency to bear.

In his book 2008 book, *Outliers,* Malcolm Gladwell famously highlighted research that showed that a disproportionate number of elite hockey players in Canada, which has an eligibility cutoff date of January 1 for age-class hockey teams, were born early in the calendar year, which meant they could be almost a year older than kids on the same team with birthdays in much later months. Older children are simply stronger, bigger, and more coordinated than younger children. Hockey coaches noticed them, gave them more attention, and put them on teams with other equally strong players, and a virtuous cycle evolved, in which these children benefited from all the positive responses that their already considerable physical advantages gave them. Gladwell pointed to research that showed this phenomenon applied to academics, as well—that older children in fourth grade scored significantly higher on key standardized tests than younger peers in the same grade.

The body of research on a child's birthday and its relationship to school-start cutoffs is expansive, with that factor often found to be consequential for many aspects of a child's life beyond sports and even grades. Kids who are old for their grade are more likely to test well, some research suggests, but also to take on leadership roles in high school. Children who start school on the young side of their grade have been found more likely to experience, in the case of females, teen pregnancy; the younger children in a grade are more likely to be diagnosed with ADHD and receive medication for it (one concern being that perhaps they are being misdiagnosed and are simply developmentally a few months behind their peers as a reflection of their younger age). Other research has found that the younger children in a grade are more likely to be judged at risk for other mental health issues or, as one paper published by the Organisation for Economic Cooperation and Development found, to feel less competent than their peers. There is nothing specific about having a birthday that falls within a particular month or two that would put someone at higher risk for actual mental health struggles. But going through school with the disadvantage of youth, relative to older peers—playing catch-up, feeling a sense of failure, when the playing field was never level to begin with—possibly does.

In the long term, the impact of school-start age is murkier: Sandra Black's research on school-starting ages even suggests that when children start school earlier, they see small gains in IQ, perhaps because older children force them to play catch-up. Other research from two Harvard economists, Susan Dynarski and David Deming, cites the trend of parents deciding to hold their boys back a year as a factor contributing to a troubling phenomenon on the rise—lower rates of high school graduation among boys than girls. The advantages and disadvantages of school-start dates may play out differently, depending on the socioeconomic status of the children who are subject to them.

But the research on the more immediate impacts of a child's age at the start of school is so well accepted by now that it has been used to shed light on the topic of how one sibling's success at school influences another's. For a long time, sociologists have noticed that when older students start to improve academically, their younger siblings usually do as well. They struggle to identify, however, what factors are actually behind that shift in the younger siblings' academic success. Has the family environment changed to be favorable for both of them? How much might their genetic similarity help explain their similar academic trajectory?

Emma Zang, a sociology professor at Yale, realized that she could look at the random element of children's birthdays relative to their school-start dates to help tease out cause and effect on siblings. Zang looked at school-start data in the state of North Carolina for thousands of students who entered school from 1988 to 2003 and came up with three meaningful findings. As expected, she found that students who were born soon after the kindergarten cutoff date of October 16— which meant they had to wait almost a full year before starting—fared better academically than those who were among the youngest in the class once they reached middle school. But she also found that the younger siblings of those old-for-their grade students benefited from their older siblings' boost—they showed stronger results academically as well, compared to the siblings of children who entered school on the young side, regardless of when their own birthdays were. And, finally, Zang discerned, the effect was larger in siblings attending high-poverty schools.

How, exactly, the influence of success in school trickles down to the younger sibling is a matter of theory. Are the older sibling's positive feelings about school somehow contagious? Does the older sibling create a more enriching intellectual environment in the home? Does she sit her sibling down and show her what she's learned? Or is it, as Rose Kennedy thought, about setting expectations?

The research on school-start dates reminds us how much in a child's life can turn on a bit of luck; it's a factor that is wholly out of the hands of that child, a random slotting that is potentially more consequential, with farther-reaching aftershocks, than the best efforts of even the most caring parents. So much of that difference is also economic: When working parents must pay for childcare, the desire to enroll a child in kindergarten in a timely fashion, may understandably supersede whatever vague concerns they have about that child being on the younger side in the classroom.

One can imagine how being young for one's class plays out over time. The child might start out as a seemingly unusually overactive kindergarten student, a child who is taxing for the teacher. That friction may translate, for the child, to a general sense of inferiority, or alienation from the one institution intended to help children learn, grow, and thrive. "All these findings together," wrote Patrick Puhani, an economist at the Leibniz University Hannover, summarizing the general theory of these findings, "suggest that early differences in maturity can propagate through the human capital accumulation process into later life and may have important implications for adult outcomes and productivity."

Even the effect of school-start dates is, of course, a result that plays out statistically, that deals with averages. There is no one choice, or even a recurring series of choices, that parents can make to protect their children from struggle or to assure them success. It is both a fantasy and a responsibility of parenthood to believe in one's power to shape a child's character; learning how to recognize the limits of one's power while taking seriously that responsibility is a large part of the project of child-rearing.

Coping with the profound role of chance, the wrench in the works, is also part of that project: For every star athlete who sprains a shoulder in the middle of recruiting season for college athletics, there's another who is lucky enough to slip into that young person's spot. At Swarthmore, when she tried out for the swim team, Marilyn Holifield had been told that another swimmer was awarded the last spot because she, unlike Marilyn, could dive; but what if she had made the team, after all, if one fewer swimmer had tried out? Would her entire experience at Swarthmore, and possibly beyond, have felt different if she'd been a thriving member of a team? A toss of the coin did not seem to change Mary Murguía's inexorable trajectory toward public service at a very high level. But if the coin toss had gone a different way, how closely would Mary have followed her sister's career trajectory, job for job for job? Would the barely perceptible differences in their personalities have yielded notable differences over time? Would she still somehow have ended up serving as the chief judge of the U.S. Court of Appeals for the Ninth Circuit?

Kathryn Paige Harden, a developmental behavioral geneticist at UT Austin, and Angela Duckworth, a relative behaviorist, share a fascination with a particular study, one that was not conducted on humans but that still captures just how much small, haphazard factors of chance early on can trigger profound, lasting results.

Imagine, if you will, forty cohabitating female mice, each four weeks old, adolescents just on the cusp of adulthood in a mouse's life cycle. Were they to gaze upon each other, and have enough higher-level reasoning, they would recognize a bevy of replicas of themselves. Bred to be genetically identical, they would prove ideal experimental subjects for a study in individualism.

At the outset of the experiment, the results of which were published in 2013, their scientific overlords at a major research institute in Dresden, Germany, placed them in a cage that was both pleasure palace and panopticon. Five stories high, it was rich with plastic and cardboard tubes, wooden scaffolds, and mouse toys—for example, hollow wicker balls ideal for nibbling, nesting, or playing. Each mouse was tagged with a transponder. Twenty antennae placed around the structure would register the mice's movements.

The mice were genetically the same; even so, the researchers knew, they would not amble along precisely the same paths through the same tubes to the same water bottles or to the same balls. They would chart different paths. Would those early choices, or happenstances, have lasting effects down the road? At the end of three months, the researchers wondered, would the mice, for all their genetic sameness, be noticeably different from one another in any significant ways? "What happens when genetically identical mice inhabit the same environment?" they asked in the paper they eventually published in the journal *Science*. How much individuality would develop over time—and could any of that individuality be observed in the very way that the brain's nervous tissue would grow?

At the end of the three months, the researchers observed a magnification of differences over time that was, in fact, "massive." Mice that had, for whatever reason, made more ambitious explorations of the cage early on continued to do so, seeking out toys and adventure; those that had, for whatever reason, remained more loyal to a few chosen spots, clung to the safe and the known over time. The researchers were even able to detect those differences at the level of neurons, showing a greater proliferation of new brain cells in the hippocampus, the part of the brain thought crucial for flexibility, in those mice that had wandered farther and sought more novelty at the outset and that continued to do so.

Scientists are loath to anthropomorphize mice, and the story of the pleasure palace–panopticon holds unknowable mysteries: Why should one mouse have been, at the starting point, more comforted or more nervous or felt more encouraged to explore farther? They were clones, after all. The metaphors are tempting but go nowhere; there is only randomness in why one mouse might have hit three fun wheels that first day and another hit only one. Perhaps one was handled more roughly as she was placed in the cage or suffered undetectable prenatal trauma while in utero? "Small perturbations," the authors wrote, "may lead to initial individual differences in action tendencies. These differences, in turn, may trigger differences in experience that accumulate over time," resulting in different developmental trajectories. Small effects—even where the mouse was first placed in the cage—could have triggered what the researchers called "a massive magnification of individual differences."

For the researchers in Germany, the finding was significant in that it showed the importance of the plasticity of the hippocampus in an organism's ability to adapt to novelty and complexity. Researchers also noted that the creation of new brain cells in that brain region could, in fact, be shaped by the way that organism interacts with its environment—that the intake of stimulation would yield capacity for more stimulation, more novelty. The more one of the cloned mice engaged with the complicated, enriching environment around it, the more that mouse may have enhanced its own capacity to learn. They were looking for insights into how the brain works, trying to build a theory of brain architecture and potential for growth. But along the way they found an elegant demonstration of the interplay of genes and the environment. "This shows it's not the genes alone, nor the environment alone. And the same is true for humans," Andreas Brandmaier, a senior research scientist at the Max Planck Institute for Human Development, and one of the researchers who published that study, explained in an interview on the digital platform BOLD, which highlights research on learning and development. "Genes and the environment interact to produce differences between one human and another, but development is a third factor that shapes the life course."

Harden also found the study illustrative of important developmental pathways. "It's not genes or their rearing environment that make the difference," said Harden of the mice in this experiment. "It's this path-dependent response to initial serendipity or randomness. My view of personality development is so much like that—it's like there are these initial deviations that might be luck or noise but that make you slightly more likely to experience something differently the next time—and eventually you get this expression of individuality."

Given the millions of ways that parents' genes can be shuffled, and given the multitude of small, random perturbations that transpire in individuals' lives, the near-infinite ways those factors can combine, a parent trying to exert any kind of influence on their children can start to feel paralyzed: Every parenting choice seems potentially important and also, quite possibly, wholly inconsequential in the face of all the other forces bearing down on children every day of their lives.

As a parent, I choose to try to see the infinite variation of possibility as not so much paralyzing as dizzying—it's liberating but also thrilling. Through simply loving our children, we have chance after chance after chance of doing something, knowingly or not, that might eventually take them down a road whose every turn we can't see but that leads somewhere beautiful.

SEVEN

The Chens

T he Chen family's journey to America is a story that begins with flight.

In one of her earliest memories, Elizabeth—she asked to be called by her American name—is standing on a moving train for hours, barely able to stay upright, a crush of people all around her. She was four; she remembers little of the great escape that train ride represented, but she knows that if they had not made it, she would have one fewer brother in her life.

Her mother, Ying, remembers all of it, like a film that unspools in her mind if she lets it, scenes in saturated color, lurid with klieg lights, alarms sounding loudly. On a warm afternoon in 1987, the phone rang: a call from a friend who worked at the police station five minutes from her home, telling her that local authorities were on their way over.

Ying was four months pregnant with her third child, in flagrant violation of a one-child policy that China had started rolling out in 1979. Even under that policy, the government sometimes made allowances for a second child, especially if the first was a daughter; but a woman carrying a third child risked being subject to a coerced abortion. Ying grabbed her son, Yi, then two, with one hand and told her husband, Xian, to carry Elizabeth. Ying threw a bag over her shoulder, and the family fled to the home of a friend who lived nearby. When it was dark enough outside for them to feel safe to leave, they approached a bus stop in front of the house. The buses there left in a steady stream, and they boarded one quickly that would take them to the train station. Sweating in the packed, hot space, Ying could feel the dissonance between her own inner panic and the calmness, the lucky obliviousness, of all those other people on their way to work or home. When the bus arrived at the station, she thought she'd be able to relax into the press of people, to disappear in the crowd, then step on a train that would carry them all to relatives in Fuqing. But as soon as she was inside the station, she heard the sirens of the officials' cars pulling up to the building. They had already searched her home, she feared, and anticipated the family's path of flight.

The train was not leaving for another thirty minutes. While her husband and Elizabeth went to get tickets, Ying, with Yi in tow, rushed into a room in the train station that was set aside for nursing mothers. The top half of the door in the nursing room was made of glass, and Ying, her pregnancy just becoming visible, crouched right beside the door, hiding behind the lower part, which was solid wood. At one point, she could hear the police on the other side of the door; she held her breath as she imagined them scanning the room through the glass. After they left, she bolted with Yi, rushing to join her husband and daughter to board the train. But even on the train, she imagined that she could feel the police searching for her, their eyes alert for a pregnant woman with a child. She took Yi with her to hide in the train's bathroom and locked the door. Eventually, when the train started moving, her clenched muscles loosened with the rhythm of its movement. She had done it: The next stop was not for hours, and she was almost sure the officials would not have remained on board.

Slowly, cautiously, she pushed her way through the crowded cars, finally reuniting with the other members of her family.

Elizabeth remembers waking up in Fuqing, where an uncle lived. They had traveled some eleven hours by a slow passenger train, and now she was in a bedroom new to her; she remembers the next several weeks as a time apart, a time when she was far from home, far from her preschool, a city girl playing outside in the field for hours.

Ying stayed in Fuqing another five months, finally giving birth to Gang late that fall, then eventually returning with her husband and two older children to her hometown, Sanming, with its cool lakes and hulking, dramatic caves. But when she returned home, it was without Gang. In the eyes of the state, her family had too many children; in the opinion of her extended family, her husband's older brother had too few, as he had none. His wife had not yet conceived. Family pressure, the threat of the state—it was all too much for Ying to overcome, and reluctant, grieving, she agreed to leave her third child with her husband's brother and his wife.

That was the second time she worried she would lose one of her children forever; soon, there would be a third.

SOME FAMILIES TELL THEIR MOST dramatic stories over and over again, softening them into something that can entertain as much as explain; some families would rather not put the past into words. Sometimes parents offer only spare details about certain parts of their own past lives to their children, setting up a force field around a painful subject, so that children mistake their own perennial fuzziness on the details for their own lack of curiosity.

Yi, Elizabeth's younger brother, was a married man of thirty-four before he heard all the details of how his mother came to arrive home without his brother Gang (pronounced "gong"; Yi is pronounced like the letter *E*). Yi was serving as the translator for me as I tried to understand his mother's past; as we pressed for more details, he also became the interviewer, suddenly curious, as his mother answered detailed questions about that time in their lives. How old was Gang when the family returned to their home in Sanming without him? (Fifty days old.) How did she remember precisely how many days it was? (Because she clung to that number, mourned it, for months after she left him behind.)

Yi, as an adult, does not see his mother often, but that week, just a few days before Thanksgiving, his parents had made the twelve-hour drive up from their home in Virginia, where they own a restaurant, to Watertown, Massachusetts, where Yi and his wife, Corisa Lee, now lived. As they talked, a downpour of icy rain outside was overpowering what had been intermittent bouts of soft snow. Ying, then age fifty-eight, wore a white fuzzy Santa hat and a white down vest and was constantly replenishing cups of tea; Xian, her husband, stayed close to the stove, shredding cabbage into small mountains for the family's dinner. He occasionally took a break from his work to answer his cell phone, explaining to customer after customer, back in Virginia, that he could not offer take-out that day, as he was visiting his children. This break for the holiday was unusual, almost unheard of for Xian and Ying, who had worked at their restaurant, the Shanghai Chinese Restaurant, for decades, seven days a week.

As Yi talked with his mother about the mad dash he had made with her, his father, and his older sister so many years earlier, Gang, then thirty years old, was also in the kitchen, along with his wife, Stephanie, both of them keeping a close eye on some high-end beef on the stove. A fourth sibling, the youngest—Devon, the only one born in the United States—had been dispatched on an errand and called Gang for some counsel as he tried to find his way back to the apartment. "What do you mean you're lost?" Gang said to Devon, who was eighteen. "You're two blocks away!"

They were all a long way from home. After immigrating to the

United States, the Chens raised their children in a city in Appalachia known as Bristol, Virginia. Ying's mother lived with them as well, as did, for many years, a cousin four years younger than Gang. For much of the siblings' childhoods, their home was so short on space that Ying and Xian slept on a sectional couch in the living room. Now, Ying and Xian had moved to a larger loft-like apartment space next to the restaurant, and all their children lived out of state. Yi and Elizabeth, a doctor, were living in the Boston area; Gang was working in San Francisco, and Devon was in the middle of his first year at Dartmouth.

Within a few years of that meal, Yi, a person who sometimes worried, as a child, about how his parents were managing the restaurant's finances, would enter a rarefied realm of the affluent: Toast, a restaurant-management technology business that he had joined as its fifth employee, went public in 2021, in the single biggest IPO in Boston's history. Its three founders, MSNBC announced, were all billionaires on paper; Yi, someone who qualified for free lunch as a child, now had enough wealth to sustain his offspring for generations to come. Gang followed his brother into the field of tech, and in 2019, after working in the field for several years, joined an English-instruction start-up, Speak, a company in a strong position to dominate the language-instruction space, with funding from the influential technology team at OpenAI.

All three of the older siblings (Devon is still a recent college graduate) have found themselves in careers and lives that empower them financially—but perhaps even more meaningfully, all three have found their success in work that could be invaluable for people facing the challenges their own parents faced for years. Elizabeth is a family medicine doctor with a sizable immigrant patient base, many of them Chinese; and both Gang and Yi went into the kind of businesses that provide services that would have eased their own parents' struggles immeasurably, had they had access to them thirty years ago. Ying and Xian's challenges at the restaurant would have been easier with the software and hardware developed with their kind of business in mind; and everything—including their parents' relationships with their children—might have been better had they been able to work on their English on their own time. It was too late to go back and ease their parents' decades-long trials; but Elizabeth, Yi, and Gang all found their paths forward by remembering the struggles of their own and their parents' past.

BEFORE ANY OF THAT COULD HAPPEN—before the family arrived in the United States—Gang remained in the home of his uncle and aunt in Fuqing. After about five months, word made it back to Ying, through another family member, that Gang had been neglected and was so malnourished that his life was at risk. Ying's mother and an uncle, with speed and certainty, managed to extract Gang from his adoptive family and return him to Ying and Xian.

The reunion of Gang with his family was a profound relief but also a crisis, perhaps the most consequential of the family's lives. As a family with three children, they would be relegated to an underclass, subject to financially crippling fines and other possible penalties. Parents with three children were judged by strangers, unwelcome in some workplaces; their children were unlikely to gain admission to university. Even registering Gang for kindergarten was a potential challenge, as they had never registered him in the city, for fear of triggering those fines. But of those who wanted a way out of China, the Chens were among the cosmically fortunate—part of the one hundredth of 1 percent, by some estimates—who had a way out, a path to the United States. A relative of Xian's was a fisherman who had finagled a sojourn in Hawaii and never got back on the boat, eventually taking advantage of a period of amnesty to win citizenship. Other family members followed, including Xian's father, which meant that now the Chens, as close relations, had a route to the United States. In 1988, when Gang was about a year old, Xian followed his father's path to the United States, planning to work and save money before bringing over the rest of the family.

In the years after her husband left, while she was raising her three children, Ying contemplated the family's future and considered some of the certainties in her life. Certainty number one: She believed that she had it in her to survive in the United States, to start a Chinese restaurant, as so many immigrants had before them, even if she had no experience in that field, few savings, and little English. In her own family, she was the child with the most common sense, the one her father relied on to help manage the family business, a welding factory that at a different point in time, before the Chinese government assumed property rights, had offered her family financial security. She was also a strong student, but her father wanted her at home to help with the business and to do the laundry and the cooking for her three brothers. She was desperate to finish high school, but her father threw away her backpack, even burned her books. Her teachers would come knock on the home door: *Ying is so talented,* they told him. *She belongs in school.* Ying knew she was bright, even if she hadn't had the schooling she wanted so badly.

As resourceful as she was, she was also certain that she could never manage to help her husband run a new restaurant in America while simultaneously trying to take care of three children. And yet she felt she could never travel across an ocean to live somewhere without them.

In the end, it was the last certainty that gave way. There was no other path; to start a restaurant and put away enough money to move forward, both she and her husband had to staff the restaurant full time. Other people could watch her children for a few years—the practice was common for families in China trying to get a financial foothold—but only she could help her husband make a living for their family.

"It was painful," she told me, as Yi translated. "But it was the only way. It wasn't that we were chasing the American dream of a perfect life—it was that we were at such a disadvantage in China. There was no hope there."

In the fall of 1992, she arranged for each of her children to live in a different home, rather than overwhelm her mother with the care of all three. Elizabeth, the oldest, then eight, stayed with a surgeon whose nephew was working at Ying's father's factory. The surgeon had three daughters. "My daughter was so strong-willed," said Ying. "I thought they'd be good role models." She placed Yi, the middle child, with a teacher whom she admired; she hoped that the teacher would discipline him, but also that others would treat Yi well because he was living in the home of a figure of high status in their community. Gang, then five, stayed in the home he had been in all along, cared for by his grandmother, a tiny, gentle woman so different from her daughter, Ying, who would prove to be fierce, a fixer, a fighter—an overcomer.

On November 11 of that year, four years after her husband had left for the United States, Ying left Sanming and took a train to Fuzhou, where she lingered for three days. "I knew I would never leave my home if I was going straight to America," she recalled. Instead, she told herself she was only going to Fuzhou. Once there, already separated from her children, she wept in her hotel room. She reasoned with herself. She grieved. She railed against injustice, the power of the state. She let herself dream a little bit about her children, who would be professionals one day, educated, cultured, and admired. She steeled herself. And then she got on a plane and left for the future.

How, exactly, do you open a restaurant—order menus, buy groceries, hire employees, manage paperwork, order and install a lit storefront sign, negotiate rent—when you do not speak the local language? To start, you do not open a restaurant right away; you become part of a system of replication, aided by other, earlier replicators. If you hope to open a Chinese restaurant, first you work in someone else's restaurant, often a family member's. In Xian's case, you go work for an uncle who has started a restaurant in upstate New York, and then find a job at another restaurant in Roanoke, Virginia. Maybe you wash dishes for years, working for a man who started out washing dishes, and you live with someone else in a small bare-bones home, probably with someone else who washes dishes. Eventually, you save a little money and you learn to cook. Then you open your own restaurant; you get advice from your family networks on where to order your menus, your ingredients, your soy sauce dispensers. Where you open your new restaurant, how you negotiate with the landlord, how you actually make a living, draw customers, make a profit—that is on you.

By the time Ying joined Xian in Virginia, he had worked in Chinese restaurants for four years. Ying started busing tables at the same restaurant in Roanoke where her husband worked, a transition that she recalls as so painful that she felt it physically. She suffered from migraines and overwhelming fatigue. She couldn't understand the customers, so she couldn't help the way she wanted to; Elizabeth recalls her mother, weepy on the phone calls she made home to her children back in Sanming. Eventually, she and her husband began the work of trying to find a location where they could start their own business.

Back in Sanming, every weekend, the children's grandmother reunited the three siblings for dinner. Even still, the separation marked a dark period for Elizabeth. The treats her parents sent her from the United States went to the surgeon's daughters; she never felt she was embraced by the family as one of their own. She remembers the waves of homesickness. "I remember that feeling—like my mom's not coming back," said Elizabeth. Of the four Chen siblings, Elizabeth is the most careful, the slowest to smile; in regards to her younger siblings, she was sometimes steely but always supportive. Now a mother herself, she tears up easily whenever she speaks of the arc of her parents' lives, just as Janet and Mary and Ramón often do when they speak of their own parents' stories.

Elizabeth, a girl isolated from her family, nonetheless carried some element of prestige, which was that everyone knew her future would unfold in the United States. Elizabeth remembers being heavily prepped for the conversation she'd eventually have at the consulate. "We were coached and coached," she said. "Just to make sure we wouldn't say anything wrong, and they would give us a visa to come."

Both Elizabeth and Yi were old enough to long for the day their parents would bring them to America. Yet it seemed like it was one of those things, like growing up, that the future supposedly held but that seemed impossible to reconcile with current reality. And then one day, after a year and a half of separation, there it was, the future merging with their present, their mother bursting in the door, sobbing at the sight of all three of her children, sitting, expectant, at their grandmother's dining-room table.

"I remember her pulling us toward her," said Elizabeth. "She couldn't stop touching us, pulling us onto her lap." Ying was only angry that they had not gained more weight. "In America," she told them, "you'll drink milk, nothing but milk," and off they went, about three weeks later, visas in hand, to the big, cold airplane that would take them to another plane that would take them to their new home. They had been patient; they had had bravery forced upon them—and now they would reap what everyone agreed was the ultimate reward: life in America.

WHEN YING BROUGHT HER CHILDREN to America, she wanted them to feel powerful, even if they knew little English. They were the only Asian children in the town where Ying and Xian had started a restaurant: Chilhowie, Virginia, a speck of a spot in Appalachia, population 1,800. The town, a half an hour's drive from the Tennessee border, was best known for its local trout fishing but soon, Ying and

Xian hoped, it would be known for the excellent Chinese food she and her husband were serving at their restaurant close to the exit off Interstate 81. She could not make her children fluent overnight in their new country's language, but she could give them proud, historical names that reflected her respect for the culture of the West and her belief that her own children would do America proud.

As she prepared her children for school, she told them how she wanted them to introduce themselves when they arrived on their first day, had them practice their new names, and sent them off.

They each arrived at school nervous, confused, and overwhelmed. "I am Elizabeth," her daughter, then ten years old, told her teacher. Her mother had given her self-assured daughter the name of an English royal. "I am Washington," her older son, then eight, told his classmates. "Hi," her younger son, then seven, was instructed to say. "I am Nixon." It was 1994.

Ying knew that Nixon had opened up China to America, and to her he was a hero, someone who had saved her family from tunneling down a path of closed doors. It was true that neither Ying nor her husband had gone to college, that their parents had not, either. But their children, they were sure, in America, would have that chance—they would be valedictorians, would go to the best American universities. Strangers in a strange land, she and her husband could raise their children to be stars without any help at all, if they had to. It was one of life's great unexpected gifts that, as it turned out, they did not have to.

Chilhowie took its name from a Cherokee phrase meaning "land of many deer." The town had a main street half a block long, one stoplight, air so fresh the family could practically taste it, and now, a small hotel with a tiny restaurant attached. The owner of the hotel let Ying and Xian use the restaurant without charging rent, because his guests needed meals, and no one else was lining up to take over the empty space. They lived in a modest apartment in two-story building a five-minute drive away.

From the moment the children arrived, they started working at the restaurant whenever they were not at school. Picture two boys and a girl, none of them more than sixty pounds, all three pushing or pulling, straining to drag bags of ice back to the restaurant from the gas station nearby. At night, when all the guests were gone, together, they elaborately folded hundreds of napkins for the next evening, filled the salt and pepper shakers and the soy sauce dispensers, peeled the strings off hundreds of snow peas, neatly laid out place mats at every seat. From a young age, Yi stood on a chair to give him the height he needed to work the deep fryer. This was their new life: a town where, at first, they understood little, and a new home, half living quarters, half restaurant. They regularly curled up and closed their eyes on the restaurant's fake red leather banquettes, sleeping until their parents had cleaned up and could carry them home in the early hours of the morning after the last cleaned, dried plate had been put back on the shelf.

In China, the future had not been promising, but for Elizabeth, at least, the present had been lighter. While their father was working in the United States, all three children had shared a bed with their mother, fighting to be the ones who got to snuggle up beside her. There was tremendous relief in being reunited, but also confusion and some pain. In China, her mother had been more playful, more demonstrably loving, than the person they came to know in America. "She was definitely more stressed," said Elizabeth. "And she was exhausted all the time." Or maybe it was just that Elizabeth had outgrown the early years of childhood that her mother considered more appropriate for that affectionate behavior.

Ying's life had, in fact, been easier in some ways in China: She worked for a family business, but she did not work seven days a week, the way she did now. In Sanming, she enjoyed a brief window when she could take Elizabeth for singing and drawing classes, the kind of thing that she herself would have loved to have enjoyed as a girl. She had support from her family, help caring for the children from her mother. Until her mother joined her in the United States five years after the children arrived, she and her husband had to manage it all—the children, the business—on their own.

The cheer and affection she had once showered on her children reached them less frequently or indirectly; now she channeled it somewhere else—toward their neighbors, their potential customers. Her desire to please was obvious, her delight in each customer equally evident. If she could not express all that she wished she could in English, she could smile, take a customer's hand, pull her to a table, and immediately start pouring tea.

From the beginning, the community in Chilhowie welcomed the Chens. Volunteers stepped forward to help tutor the children in English, and on Christmas, the father of one those tutors, a man they'd never met (and whose identity they only learned of years later, in a Facebook post) baffled the Chen children by showing up in a red felt suit with a matching cap and giving each sibling the present that child most wanted.

Ying's mother was a devoted Christian, and in Appalachia, many of the opportunities available to Ying's children came in the form of religion. When the Chen parents were busy at work preparing for the post-church rush on Sundays, Catherine Wolfenbarger, a regular guest at the restaurant, would pick the children up and bring them to church. On Wednesdays, after school, she picked them up at their home and brought them to a children's youth ministry known as AWANA, which was structured a little bit like an evangelical scout group. The children could earn badges for merit, for memorizing scripture passages, each of which was a few lines long. Some of the children would memorize one a week; some would memorize three. The Chens would come in, a family friend remembered vividly, having memorized twelve or fourteen verses, lines that started, for example, like this: "But you, Bethlehem Ephrathah, though you are small among the clans of Judah . . ."

"We didn't have photographic memories," said Yi, who remembers trying to make it past ten verses every week. "I don't think we were even smarter than the other kids. But there was a competitiveness to repeat it and memorize it so we could break records and be the best. That drive came from . . . somewhere. Maybe being the underdog." He was doing it for his own reasons, his desire to outdo the others at church and also at school—which had a direct spillover effect onto his brother Gang, who only wanted to keep up with Yi. "I just thought it was fun," said Gang.

In Sanming, the Chens had been strong students, if not standouts, but their early education meant that all three arrived in the United States knowing how to drill and how to practice. One of Elizabeth's earliest memories is of herself at age three, back in Sanming, holding a chunky pencil, making her way through lettering and math problem after math problem. "At age five or six, you're coming home with a hundred math problems so the next day you can have the reward of doing well on a timed math test," she recalled. Her mother was not unique in doing that kind of drilling or an exceptionally fierce enforcer of the work; the school expected that level of effort from all the families in the community. "Then we got here, and we saw that other students were not working that hard," she said. "For us, it's easy to work hard."

Because the Chen children were new to English, they ended up with one other advantage that might have mattered: Yi, during grades two through five, and Elizabeth, for all her time in school, were among the oldest children in their respective grades, because the school decided they could more easily catch up if they started a grade behind. Whether they were exceptional or not, they might have had developmental skills that outshone those of other students in their classes who were a year younger—self-control, focus, attention. Elizabeth felt that teachers always singled her out right away as a talented student, which she attributed to a favorable bias she thought they might have felt toward Asian American kids. But maybe it was her relative maturity that they noticed, an advantage that fed into her younger siblings' success.

The Chen children arrived in Virginia primed with the kinds of skills that adults generally admire—and also their mother's expectations. At some point, Gang was singled out for a gifted-and-talented program, but Yi was not. "And my mom refused to believe I wasn't gifted," said Yi. Ying recalled that Yi also expressed the feeling that he was bored in class. She decided if the school was not going to enroll Yi in the gifted-and-talented program, she'd have to figure out another way to keep him engaged—so she pulled him out of fifth grade and dropped him into the middle school as a sixth grader, mid-year, where he would have to start over socially.

The contrast between the barriers that would have held them back in China and what looked to Ying like a wide-open horizon of opportunity in the United States never ceased to energize her. "In China, you could work hard and do everything right, and still you had no chance of getting ahead," said Ying, trying to explain why she expected her kids to take on the most challenging workload they could. "But in America, all you had to do was work hard in school, and you could make something of your life."

After the Chen children had spent about a year in Chilhowie, the hotel owner sold his property. By then the restaurant was thriving, and the new owner wanted to take over the space. The Chens spent the next year living off whatever Xian could make working in other people's restaurants, and Ying took on factory work. Both parents spent hours driving around Virginia, looking for a new place to start over. On those days when their parents were on the road, the children went straight home after the bus dropped them off. Often their parents did not come home until long after dark, which made them feel uneasy. When one of them showered, the other two stood guard outside the apartment bathroom with bats, just in case. Their mother left dinner for them, sometimes just plain rice; money was so tight that when they once ordered in pizza for a change of the routine, their mother, angry, reminded them that they had no money for that kind of extravagance.

But even in that first year the children were in the United States, the children's home held a piano, rickety, but functional, a donation from a kind neighbor. Before they went to sleep, they each made sure to practice piano. Because they knew that their mother's appraisal awaited them the next morning, before they fell asleep in a bedroom all three of them shared, they took turns playing, filling the small, spare space with tentative, repetitive notes until they put themselves to bed.

EVENTUALLY, THE CHENS FOUND a restaurant space in Bristol, a town a thirty-minute drive from Chilhowie, with its own NASCAR track ("the fastest half mile in the country") and a country music museum. Once a railroad hub with a booming coal industry, Bristol was sliding into decline by the time the Chens moved there in 1996. By 2002, the poverty rate there was 17 percent, almost twice as high as the average poverty rate in Virginia as a whole.

About a seven-minute drive away from the restaurant, they found a home, a small red-brick bullet of a building, sturdy, its windows small and spare, its yard large enough that Ying could coax a garden of Chinese vegetables to grow. The restaurant itself was perfectly situated, in that it was adjacent to the town library, and the Chens could leave their children alone there, safe among the books, while they worked. The Shanghai Chinese Restaurant, a Bristol institution, is still in operation today, although it moved to the structure next door.

A large part of the restaurant's décor consists of family photos: Devon serving as drum major for the high school marching band; Elizabeth, glamorous in a sleeveless white gown, at her wedding (to someone who, like her, immigrated from China when he was a child); Devon, small and chunky, at around age seven in a green suit. Another image shows Yi, buff in a sleeveless shirt, beside Gang, smiling, with Devon in his arms; it's close to one that captures the whole family on the first vacation they had taken in about ten years, a trip to China (which Elizabeth visited again, with her maternal grandmother, before heading to college).

The Shanghai Chinese Restaurant was home for the Chens, and it was the place that helped make it possible for them to call Bristol home. After school, the children came straight to the restaurant and sat, all three of them, at a table, doing their homework. They were on display, as if they were living their lives—demonstrating their industry—in public, a play within the play that was the life of the restaurant. The restaurant work of the family was also always on display, the kitchen open to view. The customers would see that, see how much it took out of the family to keep it running. When Devon, born when Elizabeth was sixteen, was still a baby, Cathy Wolfenbarger remembers seeing Ying, exhausted, take a nap in a playpen, right there beside him, just a few feet from the cash register. "We all saw just how hard they worked," she said, "and it made you want to support them." Ying was cheerful, she was unfailingly friendly to customers, but she could not disguise how bone-weary she and her husband were.

For the Chens, running a restaurant would turn out to be a remarkably powerful vehicle of belonging. If you run a shoe store, maybe you see the bulk of your customers once a year, in the fall, a time to comment on how much everyone has grown, another way of saying how much time has passed since you've last seen them. If you own a restaurant, particularly in a small town, and if you serve a crowd-pleasing Sunday buffet, almost everyone passes through, if not every week, then often enough. You are not selling them shoes; you are sharing with them your food, your culture. You are selling them comfort, pleasure, communion. In the case of the Chens, the family was also, consciously or not, selling something else: the Chens.

"They were the most warm people," recalled Wolfenbarger. "What they'll do is, they'll come over and sit down with you while you eat—this happened many, many times. If I had a bad cold, Ying would give me something, and say, 'Here, you need to take this.' She used to tell me every time I came in how beautiful I was." The Chen children, too, charmed her. "There was not a single, solitary time when they were not gracious and mannerly and kind," she said.

Because of their presence at the restaurant, whatever made the Chen children exceptional would not go unnoticed. In an under-resourced area like Bristol, the people who had enough money to enjoy restaurants tended to be more financially stable—they were, in other words, in a position to help. For their first year or so in the United States, the children practiced on that donated, rickety piano—a gift that would prove momentous.

In the coming years, music would become a joy, an advantage, and a launching pad in each of their lives; it was also a source of some of the most anguished moments in their home.

Ying's father had played the erhu, a two-stringed bow instrument, beautifully; Ying, too, loved music but had been denied lessons. There seemed to be a direct correlation between how angry she was about being denied various opportunities and how much she insisted her children take advantage of the ones in front of them; both were extreme.

Many parents would like their small children to practice an instrument for even fifteen minutes at a time; very few are ready, typically, to do much more than that, and act out or refuse. What kept Gang and his siblings from refusing, beyond their early training?

"Fear of death?" Gang said, only half joking. It was true that they were unusually disciplined children, but even they might have drifted away from extended practice—at least an hour was expected, even when they were very young—had it not been, early on in their musical careers, for their mother's insistence. Apparently, there were only a few things that set their mother into an unnerving, unchecked rage. "One," said Gang, "was if she felt like she was being taken advantage of. The other was when we didn't practice." Without adequate training, she could not sit down beside her children and guide them or refine their technique; what she could do was use force and fear to ensure that they practiced.

It would be easy to fall back on familiar narratives that would explain why Ying made that effort and applied that kind of pressure—a path for her children to the Ivy League, the gratification of seeing her children celebrated in public. But Ying worked such long hours at the restaurant that she often missed the performances at which her children would eventually solo; she was working those hours to pay for the lessons and instruments, rather than hiring help that would give her the occasional night off. She was not someone who was trying to maintain some family status or keep up with the high-achieving neighbors; when the first three children were young, she had little understanding, her older children say, of the kinds of extracurriculars that would gain a child in the United States acceptance to a top school, beyond stellar grades.

Ying's children do not idealize their mother, but they do resist oversimplifications of what motivated her. "I think it really was the idea of bringing beauty into our lives," Elizabeth said. "It wasn't until later that she saw how many doors it opened." Yi believes his mother's relationship to their musical engagement was more complex than a simple matter of pride or ambition—he sensed that she saw they had potential, and that she wanted them to enjoy the richness of what music could offer them.

When I put the question to Ying directly, she told me that she felt so passionately about music because she'd seen what it had meant to her father, who had suffered so much misfortunate over the course of his life: He lost his father as a child, received little education, built a thriving business in spite of that—then lost that business during the Cultural Revolution that started in 1966, a reversal that impoverished him. "He had a lot of hardship," she told me. "But he was a wonderful musician who played various Chinese instruments—string instruments, percussion instruments. And he was a happy man. I thought as long as my children were skilled musicians, they'd be happy—music would open their hearts and minds."

Like Amalia Murguía's benediction prayer, she hoped, music would give her children a superpower that transcended financial status. If her children could play musical instruments, she told me, they would always have a place of community, of actual harmony.

Whatever her motives, however well-intentioned, her management of their musical training was punishing: It wasn't unusual for her to hit her children when she was in one of those rages, sometimes grabbing whatever was handy—a pair of tongs, a spatula, or a belt. As children, they genuinely feared their mother. Her insults were harsh. "Once, I teared up, just hearing it," said Angela Thorp, a woman the school had enlisted to tutor the children in English soon after they arrived.

Over the course of her children's lives, Ying would induce tremendous amounts of anxiety in all four of them, all because she was deeply convinced that music would make their lives better, more beautiful.

Now a parent himself, Gang is appalled by how his mother disciplined her children; he doesn't excuse her physical attacks, but he does have more insights into the origins of her rage, how she might have perceived their everyday distractions as an offense to all that she had done to put opportunities in front of them. He recognizes that his mother, especially in her earliest years in the United States, had so little support in the face of those emotions: "No money, no community, no friends." Ying, he said, could force her children into compliance and practice, but her unyielding fierceness also damaged many relationships, including the ones she had with the children on whose behalf she was working so hard.

Elizabeth, whose complicated feelings about her mother have endured, shares Gang's opinion on corporal punishment; she, too, has compassion for her mother, who was raising her children with the same kind of discipline she had likely experienced herself.

Their father, Xian, was a much calmer figure in their lives, someone who they sensed sympathized with them, even about some of their mother's demands. "Spiritually, I'm closer to him," said Gang. I asked Ying if she felt that the children benefited from having one parent who didn't put pressure on them to achieve. "No, it would have been better if he'd pushed, too," she said.

THE RESTAURANT WAS THE CHEN family's livelihood, but it was also their home, the place where the children spent their waking hours when they weren't at school. Their parents kept a piano in a backroom of the restaurant, and later, moved a baby grand into the dining room. As diners sat and contemplated specialties like Xian's thin-skinned egg rolls or roast Peking duck, they could often hear one of the Chen children practicing. Maybe one was plunking away, first at a Czerny exercise, then, many months later, at some simple Mozart minuets, possibly going over and over the same four bars, hitting the same wrong notes—D-flat, D-flat, D-flat—before finally getting that reach: ah, the resolution of E. Sometimes, when they were older and played more challenging pieces—something showy, Beethoven's *Moonlight Sonata*, maybe—at the completion of the piece, guests at several tables would invariably burst into applause. In middle school, each of the children had picked up another instrument—for Elizabeth it was clarinet; for Yi, trumpet; for Gang, saxophone—and customers sometimes dined to the accompaniment of those instruments as well.

Often, as one of the children was practicing piano, a restaurant guest would wander over, put a tip in the jar that they kept on the top of the piano. "Sometimes even when it was clear we were just practicing, they tipped us anyway," said Gang. "I'm sure we sounded awful, and it must have been very annoying." But those customers were saying something more important than "thank you"; they were saying, *Keep going.*

The restaurant never provided the Chens with much in the way of money—year after year, the margins remained perpetually small—but they turned it into a resource; it was not just a restaurant, but a highly enriching environment for their children. Their customers were varied—the heating systems repair guy, the NASCAR staff, the retir-ees, and the brewpub bartenders—but it was also heavy on doctors, lawyers, and music teachers, professionals who brought their friends and who told Elizabeth where in the area she could go to hear high-quality classical music, what events not to miss.

Elizabeth often sat down with some of her favorite customers, kept them company as they ate dinner, took in whatever advice they had to offer. They were parents of friends of hers from school, or just people who liked Chinese food—people who, when the time came, told her where they thought she should apply to college, stepping in for an overwhelmed guidance counselor at the high school.

Many of the children's teachers came to the restaurant to support the family and found themselves being supported instead. The chil-dren's middle school band director, Ben Parks, took his lunch break there every day, along with a colleague. "I offered every time to pay," he said. "Ying wouldn't hear of it."

For Ying, serving the teachers was a way of showing respect but also cementing professional musicians' bonds with the family. Ben Parks came regularly, but also started bringing his friends, other peo-ple who liked music. "I remember going there once and thinking, *What is this, a woodwinds convention?*" recalled Tom Crawford, one of the children's music teachers, who said he always ran into musicians he knew there, people drawn by what they had heard about the chil-dren. Another customer ran the marching band at a nearby high school and ended up volunteering to give Devon lessons so he could go on to become his own high school marching band's drum major. The children's fifth-grade teacher, Ms. Brittle, who later became a school principal, also dined at the restaurant frequently. She often brought Elizabeth books: *Memoirs of a Geisha* and *Snow Falling on Cedars*. Another customer brought her *Gates of Fire*, a best-selling his-torical novel about the three hundred Spartan warriors who led an alliance of Greek fighters against a Persian army that vastly outnum-bered them.

Elizabeth read that book three times, inspired by the Spartans' discipline, perhaps not even recognizing that it reflected back to her some of the discipline she saw in her parents, who worked six or seven days a week. "The hardship of the exercises is intended less to strengthen the back than to toughen the mind," pronounced one great teacher in the book. "The Spartans say that any army may win while it still has its legs under it; the real test comes when all strength is fled and the men must produce victory on will alone." Also: "Habit will be your champion. When you train the mind to think one way and one way only, when you refuse to allow it to think in another, that will produce great strength in battle." The book left an impression on Elizabeth, who would go on to try to imprint those kinds of impressions on her brothers. "And I remember thinking, *If these people can train to be warriors like that, my life's not that hard,*" she said. "*I can work so much harder to achieve my goals.*"

If there were superb educators in the area, no matter how tenuous the connection, Ying managed to persuade them to help her children. Devon remembers going on Sundays to a retirement community where he received music theory lessons from Mary Landrum, an otherwise retired music professor in her eighties, a pianist and organist who'd received, as a young woman, a master's degree from the Eastman School of Music. Ms. Landrum offered her deep knowledge of music history as a gift to her young student; in return, at the end of the lesson, Ying sat beside her, spoon-feeding her soup or fried rice she'd brought with her from the restaurant.

His mother seized on opportunities where she saw them, took every chance she could for her children to collide with talent. Gang was in the middle of band rehearsal one afternoon when he got a call on his cell phone from his mother: A renowned pianist, a woman who was performing that night at a theater in town, had just shown up at the restaurant, she told him. Ying wanted Gang, who was in seventh grade, to join them immediately so he could play for her the latest piece he had been practicing, a Chopin nocturne. Gang reminded her that he wasn't due at the restaurant anytime soon, told her a firm no, returned to his friends and to practice, felt a moment of relief, and then forgot all about it. But the next evening, when he showed up at the restaurant for work, he immediately saw a woman whom he did not recognize. He could tell right away—something about the geometry of her haircut, the fineness of her coat—that she was not from far southwestern Virginia. His mother brought him over to introduce him: It was the pianist, who had come back, at his mother's insistence, to hear him play. Gang pulled his mother into a small room in the back of the restaurant, at which point the two embarked on a screaming match that was all the more dramatic because they both knew their esteemed guest outside could likely hear every word of it.

"Why do you have to embarrass me like this?" Gang shouted (as best as he recalls). "She doesn't want to hear me play. I'm not even ready! I'm not this person you can just march out to perform!" His mother yelled back that he had no choice—she had already told this woman he would play; now she was waiting. Perhaps the esteemed guest even heard the sound of a scuffle—Gang's mother physically dragging him toward the dining room, Gang resisting. At last, the two of them emerged, Ying smiling, Gang red-faced, unable to make eye contact. "In the end, what do you expect?" said Gang. "She's my mom. So I ended up playing for her."

Gang was playing Chopin's "Nocturne in C-sharp Minor," a piece with a haunting, leisurely left hand and a more insistent, trilling right hand. He felt the woman's calmness as he played, a quiet charisma he associated with her accomplishments. "As much as you love playing the left hand, keep it quiet," he recalled the pianist telling him. "Also, forget the time signature. With Chopin nocturnes you can't go with what's written on the page; it's very much you feeling the measure, you feeling the tempo."

It was Gang's first time playing for a pianist of that caliber. The intensity of the emotional memory has fused with the music, so that although the piece is now one of his favorites, he can't remember if it always was or became so after the encounter. "Did my mother just want to show me off, or was it actually about getting instruction?" Gang said. "I don't know."

As the children got older, the balance of their identities shifted away from the restaurant and toward their lives at school. By the time Elizabeth was in high school, the family dynamics had already realigned in ways that defy pat assumptions. Ying and Xian were too busy with work to micromanage their children. The family spent so little time in the house that the years went by without their kitchen having a counter above the lower cabinets, as their mother lost interest in a renovation shortly after moving forward with it. Even if they'd had the time to oversee their children's activities, the Chen parents were unfamiliar with the school system in America. "Our mother knew how to push us," said Elizabeth, "but by the time we were in high school, maybe even middle school, she didn't know the details of what classes we should take, or what we should try out for."

In 2000, at age eight, one of the children's cousins came to live with them while his parents, newly immigrated to Europe, were trying to put their lives together; at that age, he was old enough to recognize the unusual aspects of the family he had joined somewhat uneasily. The cousin—who preferred that his name not be used—was homesick for his own family and in no position to keep up with three siblings who were already, by then, stars at school. "They were too out there," he said. "There was no way I could compete."

Their cousin acknowledges that Ying had a strong personality and that she was clearly ambitious on behalf of her family, but he also admired that she was, in some ways, unconventional, different from many of his other relatives in the area, many of whom also worked in restaurants. "I had other cousins whose parents told them 'Stop wasting your time playing piano, just help out in the restaurant, we need you here,'" he said. "She definitely placed their extracurricular activities first and then the restaurant second."

Even more unusual—and more formative, he thought—than Ying was the unusual closeness of the siblings, and the way they nudged one another, in a warm and supportive way, to achieve. "The way those three pushed each other was the key to their success," he observed. They weren't geniuses, he said, but the siblings' interactions amplified whatever talents they had. When he pictures one Chen child playing piano, a sibling is on the bench as well, refining the younger sibling's technique; they leaned over homework together, the older teaching the younger.

Although Elizabeth, in high school, had three more years of training on the clarinet than Gang did on the saxophone, she was eager to play duets with him and sensed that he was not being pushed at school. "So I made him play these pieces that were much harder than what he would have just been learning at school," she said.

In high school, Elizabeth had started taking private lessons with Crawford, the future restaurant customer who was a well-respected woodwind instructor, a graduate of the San Francisco Conservatory of Music. "With Mr. Crawford, I found someone I really wanted to work hard to impress," said Elizabeth. "I was much more afraid of disappointing him than my mother. And so I made Gang start taking lessons there, too." (Crawford, in turn, found reasons to adjust hourly rates downward to make the lessons affordable for the family.)

Elizabeth became a kind of parent figure, demanding, but reasonable, and with a better grasp of what opportunities her brothers would benefit from outside the realm of music. She encouraged her brother Yi to throw his energy into wrestling, believing it was the rare sport in which someone Yi's size—he was slight compared to his peers—would not be disadvantaged. But even more than that, she thought the coach was someone whose particular style of mentoring would benefit Yi.

The siblings were superb motivators for one another; but even as a high school junior, Elizabeth had learned from her mother's example— she encouraged her siblings to seek out mentors who themselves demonstrated excellence. Mr. Crawford was one of them; the wrestling coach at the school, Ed Cressel, was another, a local legend for his high standards who had won the state championship as a young man in 1971. "It seemed to me he led a very disciplined life," said Elizabeth, "which is not something I saw in other coaches. The wrestlers felt he cared about them. I admired him."

Yi's strong suit, always, was discipline. Of all the Chen children, Yi was the most devoted to helping run the restaurant. For Elizabeth and Gang, the restaurant was a duty foisted on them, a hassle they had to manage to protect time for schoolwork and time for friends. Somehow, Yi—maybe because he was the older of the two brothers, Elizabeth and Gang thought—took the responsibility of the restaurant on as his own. From the time he was nine, he was the one tallying receipts with the calculator, staring down the numbers, and, soon after that, asking questions of his parents about how far those numbers would carry them. When he mopped the floor late at night, he was feeling the urgency of the restaurant's solvency, which made him mop with all the more fervor; he was forever trying to anticipate the broken piece of equipment, the shortage of help, before it happened. For Yi, worry was the problem and working the antidote.

The summer after his freshman year, he took the tips he earned working at the restaurant and went away to a one-week wrestling camp run by two former Olympians. His years at the restaurant, he thought, prepared him for the mental toughness of a sport that left no room for excuses or finger-pointing: It was just you and your opponent out there on the mat, with no backup. His sophomore year, he started winning. Every match was a puzzle he had to solve, not with will alone, or brute strength, but with resourcefulness—with an accumulation of skills, which he improved with drills and grappling, over and over again, with Ed Cressel, who still got onto the mat with his wrestlers to show them what a state champion looked like.

"He seemed to me like a kid who had something to prove," said Cressel. "But he also had that innate thing that good wrestlers have, that mental toughness." Cressel characterized his coaching style as "constant pounding and constant brainwashing," reinforcing, consistently, one message: *Refuse to lose.* By the time he was a senior, Yi was ranked second in the state for most of the year; he made it all the way to the semifinals before landing in sixth place in his weight category statewide. His parents came to only one match, because their schedules didn't allow for more than that. The one time his grandmother watched a match, she tried to lunge for the opponent, who seemed, to her, to be beating her grandchild to a pulp as adults and other children cheered him on.

The Chen parents had, by the time their three oldest were in high school, turned the management of their children's lives over to their teachers, instructors, and coaches—and the children themselves. Ying was distracted by a late-in-life newborn, Devon, who was born when Elizabeth was already sixteen. The Chen children formed a counsel of three, a posse, not unlike the Murguías, a small family consortium of siblings consulting one another carefully on their choices: what sports to play (Gang followed Yi's lead into wrestling), which advanced placement classes to take. But while the youngest four Murguías were fueled by competition to be the best of the group, the Chens each wanted their younger siblings to outstrip them. "We always wanted the younger sibling to do better than we had done," said Yi.

Competition among siblings is so often the painful, primal consequence of so much closeness, especially when resources—affection, or money for opportunities—are in short supply. The Chens, possibly, had been left so totally to their own devices that they circled around one another for mutual support. Elizabeth, who set the tone for her siblings, also never stopped feeling grateful for their reunion, having been apart from her brothers during that long, painful separation in their childhood.

Their shared experience in America had also bonded them as much as any values their parents passed on. "We were also the only ones who had any idea what each of us was going through," said Elizabeth. "Our experience in Bristol was unique." The Asian American population in Bristol was less than half of 1 percent; the town was 95 percent white, which made it more diverse than most of the towns nearby. In an essay Devon wrote when applying to college, he expressed great affection for his hometown; but he also recalled endless jokes about eating dog, and being asked where he was from (he was born in Bristol) or to speak Mandarin on demand.

Just as Sarah Groff surpassed her older sister as a swimmer, Gang excelled in wrestling even more quickly than Yi had. He wasn't more motivated, as Sarah was, but he did benefit from having an older brother willing and eager to devote hours to critiquing his form.

His sophomore year, Gang had to make a decision: Would he try to compete at the state finals for wrestling—his coach believed he had a shot at winning in his weight class—or would he audition for All-State Band? Coach Cressel knew his family's values were with the musical commitment, and he did not stand in Gang's way. Gang was competing with some of the most privileged students in Virginia, the children of diplomats and government employees who lived right outside Washington, DC. He knew, from conversations with friends he made at music summer camps, that their parents might be paying up to $200 an hour for lessons from renowned instructors. Gang found out two weeks after the audition: He was not only the first student in his high school's history to make it to All-State Band, but he had landed the most prestigious spot for his instrument—first chair, a victory he would claim his senior year as well.

Gang brought home the equivalent of the gold medal. But, if anything, the Chen children's unusual success can be explained by seemingly opposing forces: They had a mother who made it clear that she expected her children to excel, but they also had a mother whose time constraints were so great that she had little choice but to grant them near-total independence. They were forced to rely on one another and forced into mature roles earlier than their peers.

Sometimes they strained, emotionally, to cope with the

independence they were given at such young ages. "We had all the power," said Elizabeth, "because we knew English. And sometimes we were probably disrespectful, because we had all this power, just by knowing English." They were the ones dealing with any legal issues at the restaurant—a letter from the health department, a bill that was late—and they were the ones in charge of the deadlines for a summer-camp scholarship fund or an application for a statewide educational opportunity called Governor's School.

We are not cultured. That was what Ying and her husband would say to Gang, in Chinese, whenever he told them—with a clap of enlightenment, or sometimes embarrassment—that they might have done something a different, better way. As the children got older, they would tell their parents just how mistaken they were about schools, or dating, or how to run a Chinese restaurant, which the Chens did brilliantly or passably, depending on your standard of success. *We are not cultured*—it was another way of saying: *We are not educated.*

Elizabeth, who was valedictorian of her class, went to Vanderbilt on a Gates Millennial scholarship, which covered everything the university did not, studying pre-med and clarinet before deciding to focus on her pre-med classes. Yi (also valedictorian) attended Cornell, his costs covered by a combination of loans and a generous scholarship. When Gang (also valedictorian) was a senior in high school and Elizabeth was still at Vanderbilt, he applied to Yale but was at a wrestling match away from the internet the day acceptances went out. From a lab where she was working, Elizabeth tried to log in on his behalf, but he could not remember his PIN and kept giving her the wrong ones. "You're such an idiot, you don't deserve to get in," she yelled at him.

She was nervous, not just because she wanted the best for him but because she knew he had written his college essay about her—about the way that she had first translated America for him, told him what to wear to school, when to make sure he had money from home for lunch or for a school trip so that the teachers didn't have to reach into their own pockets. "And then later, she was trying to translate my parents to me," said Gang, "both literally and more generally. My Chinese was never that good, so if I needed to talk to them about anything complicated, I needed Elizabeth to do it. But also, she would try to help me see things from their point of view."

Elizabeth could still remember that hot evening in China when they fled to save Gang's life, and she remembered better than he ever could how much their parents had left behind for him and for Yi—her grandfather's house full of cousins, big family trips to the park on weekends, the basic feeling of not being foreigners. Elizabeth tried not to add more pressure to her brothers, but as she got older, she was insistent that her brothers reach their own potential.

As she waited for Gang to remember the right PIN, Elizabeth felt somehow like the response from this American university would be a validation not just of Gang's worth but of her own, as his guide and her parents' proxy. "Because he had written his essay about me, somehow that made me feel like if he didn't get in, it was because I wasn't good enough," said Elizabeth. "What if it wasn't good enough?"

Finally, Gang knew he had remembered the right PIN—he gave it to Elizabeth, but he heard only silence over the phone. A lab manager who happened to be there at the time made fun of her later: One minute she was screaming at her brother about what an idiot he was, and the next she was silent, with tears running down her face, as she stared at the massive word scrolling across the screen: *Congratulations.*

HAVE THE CHEN CHILDREN BEEN SUCCESSFUL, as their mother defined it? This is the problem with expectations—they hover over households, more vague and less resolvable than a list with boxes that can be checked off.

Elizabeth believes that she chose medicine as a profession because of what she witnessed as a teenager, when her mother was pregnant with Devon. At the time, the proof of her parents' romantic life felt like a personal humiliation. "I did not speak to my mother for a month," she said. But as usual, she needed to play an adult role in this next phase of her mother's life, serving as a translator between her mother and her ob-gyn, a regular customer at the restaurant from whom they never saw a medical bill. Until that time, Elizabeth had never really interacted with a doctor; none of the children went for regular checkups, a luxury for which the Chens had neither time nor money. She was bowled over, witnessing her mother, who served so many, who worked so hard, in the role as the object of someone else's careful concern. "I was really amazed by how much the doctors cared about her, not just about the pregnancy, her ability to be a mother— her own health was important," said Elizabeth.

Some years, when she was in residency, Elizabeth worked eleven-hour days, and her mother worried that her daughter was working too hard—that maybe she should have pursued music after all, rather than feeling some pressure to choose the safer option of medicine. Wasn't the point of all Ying's effort to provide for her daughter so she could follow her dreams or have a life of ease? "Now my mom will say, I regret pressuring you to go into medicine," Elizabeth told me at one point. Those comments pained Elizabeth "because it's a decision I made on my own," she said. But those comments also triggered other insecurities: Was her mother not proud of her? Was she disappointed? Did she think Yi and Gang had made better choices?

Yi, who studied electrical engineering at Cornell, started his professional life with a grueling workload as intense as the one that medical school and the early years of Elizabeth's medical practice imposed on her. He graduated with a job at General Electric, overseeing the programming of locomotive software products, which required him to log long hours on trains traveling through Canada. He was living in Orlando at the time, and he was frequently flying up to remote towns in Canada where he would wait for hours at small outposts. Once on the train, it was his job to spend as many as twenty-two hours in a row awake, making sure that the system he was overseeing was working as it should.

"I had that feeling like I used to have at the restaurant," he said, "like I have to give this everything." Exhausted and stressed, he ended up collapsing on one of the trains, ill from what turned out to be an ulcer. Yi did not like to accept limits, but he had to acknowledge that his work ethic was physically hurting him.

After taking some time to recover, he applied and was accepted to Harvard Business School. A posting on an alumni page inspired him to get in touch with Toast, which had only four employees at the time (to make themselves look more official, they met him, for his interview, at an investor's office). They wanted to sell systems to restaurants that would allow them to manage take-out orders, keep track of inventory, manage gift cards, payroll—all the small and large features that Yi knew would have made the Chens' lives easier, their workdays less stressful. He knew that even as a business student, he could be immediately valuable at Toast.

He signed on in 2013 to work in the basement of one of the founders' houses that summer and quickly made the business's first sale—to a cousin who owned a local coffee shop. One of their first eight sales: The Shanghai Chinese Restaurant of Bristol, Virginia. Yi, who helped write Toast's mission statement, used to relish reciting it: "To empower communities to do what they love: delight guests and thrive." He had given so much of himself to his family's restaurant; now, he sometimes said, it was giving back to him. Having worked at a restaurant for so long, he could intuit what services would be valuable and what bugs would be problematic. "People often ask me, 'How did you know it was going to be this giant, amazing success?'" said Yi. "I say, 'I didn't—I did it because we were going to build something that can be helpful to people.'"

Even before the pandemic, Yi was always interested in the services that restaurants could employ to build their relationships with customers, to foster the kind of bond he knew the Bristol community felt with his family's restaurant. His parents knew the coal miners who walked in, they knew the teachers, they knew the police officers, and they knew, after decades in the same location, those people's children. Was there a way he could help restaurants in bigger cities cultivate those kinds of connections? Asked what department he wanted to lead, he took on the consumer digital business: enabling online orders for takeout and delivery, loyalty points, gift cards, much of which deepened restaurants' knowledge about their customers, connecting them with their email addresses and providing more outlets for sharing information.

At the time Yi started developing that aspect of the business, it was a small portion of Toast's portfolio. Then the pandemic hit. Within a matter of months, restaurants in most urban areas lost up to 80 percent of their revenue, by some estimates. Yi's division became the aspect of the start-up that could actually serve the most urgent needs of the restaurants that remained in business. The company still ended up having to lay off 50 percent of its employees (though it soon regained that head count), but, within the leaner business, Yi's team scaled up the services that would help the company keep growing. Toast, valued at $8 billion in 2020, was valued at around $30 billion when it went public in September 2021.

Yi hesitated over whether to bring his parents to Boston for the initial public offering. Because they are still not confident in their English and because they might have trouble navigating transportation, he knew that it would add considerable stress over those hectic few days to have them there. Devon, taking a semester off from Dartmouth, was living with Yi at the time and expressed his own strong point of view. "It wasn't even my accomplishment," Devon said. "But I thought it was really important that my parents be there to see Yi's accomplishment." In the end, Yi's parents came, bore witness to the balloons falling from the ceiling, the celebratory party, the jittery energy at the stock exchange—the cheering as the shares' value soared.

Of the three oldest children, Gang was the sunniest, a bit more philosophical than his siblings, as hardworking as they were but with a gentler edge. By the time he turned thirty, he was working at a tech company in San Francisco called Thumbtack, which connected consumers with service providers. He found the job engaging and the challenges absorbing, but sometimes he thought he didn't know what he was working toward, or even who he was supposed to be. His parents had worked eighteen-hour days, almost every day of the year, for a decade, to give him choices. Now he felt that the best way to honor what they had done was not just to work hard but to take advantage of the freedom he had to reflect on where he should devote his talents.

As a child, Gang had been a gifted painter, talented enough that customers bought his work, if only to encourage him. Even when he was at Yale, where he'd chosen to study foreign policy, his mother sometimes wondered aloud to him—it was maddening, in its own way—if he should be pursuing his art instead.

Because he was the youngest of the three siblings who immigrated, Gang's Mandarin was never strong enough for him to communicate with his parents with any real nuance; when a teacher in middle school began a section on China's Cultural Revolution, he asked Gang to help lead the classroom instruction. Gang's own family had lived and suffered through it—but he knew no more about the subject than any of his fellow students. Right out of college, he had decided to move to China, to teach English in Changsha, in Hunan Province. In Changsha, he fell in love with Stephanie Cheng, a fellow Yale graduate teaching there.

Gang's parents wanted him to live the life he found most rewarding, rather than chase financial security, although as it happened, that had come to him. He was two years into his relationship with Stephanie before he visited her at her parents' home in Napa Valley. Only then did he fully understand the extent of the gap between his family upbringing and hers. "It's a little over the top," she had warned him before they visited the sprawling two-hundred-acre estate, with its vineyards, its lake, its two putting greens. Her father, originally from Hong Kong, had taken a family business in metal parts and turned it into a cookware empire.

The two were married before Stephanie had ever been to Bristol. They managed to squeeze in a visit just as his parents were moving into an apartment in a building they had bought next door to the restaurant. Just as nervous as Stephanie was when she first showed Gang her home, with its tapestries, its nineteenth-century Steinway piano, its high vaulted ceilings, he now showed her around his own. Stephanie took it all in: the chipped green counter, the second shower packed with cleaning supplies, the living room sectional that had served as the Chen parents' bed all those years. She said little but took his hand as they walked out the front door, where a mostly red and gold image of a Chinese mythical god was hung, with Chinese lettering on a scroll. It translated to *Congratulations, and may you prosper.*

When Gang looked back on his career to that point, he realized that he'd found more meaning in his work in the years right after college, when he was teaching English in Changsha; he could see how much the work he was doing directly affected the lives of the students. He still thinks about one of the students, a young man who was tracked into the lowest level of students at the school, who labored for countless hours to audition for and land the lead in the play they put on in English, *Beauty and the Beast*. The opportunity, the student told him, was one of the high points of his life, a meaningful and rare moment of recognition. Many of the students Gang taught have since immigrated to the United States or Canada, and he knows he helped ease their way. The coaches and teachers in his own life had been so meaningful to him; he realized that he wanted to be able to provide that for others.

He knew, too, how different his parents' lives would have been had they been able to speak English more fluently. They wouldn't have needed to lean so heavily on their children. They might have traveled more widely. His mother, almost certainly, would have finally gone to school, maybe even gone on to get her college degree. And his own connection to his parents might have been deeper. Discussions about feelings or complex ambiguities—all of that went silent, never expressed or even attempted. Not long after he and Steph returned from taking several months off from work to travel, he accepted a job at Speak, a language-instruction program that focused, initially, primarily on English instruction, using AI to create teaching conditions that are responsive and conversational. Its capabilities were turbocharged in 2022 when OpenAI, the AI lab that works closely with Microsoft, publicly invested in the business by way of partnership and funds. Gang's job title there: head of growth.

DEVON HAS ALWAYS EMBODIED, for the Chen family, a version of themselves that reflects the influence of an entirely American upbringing. Devon towers over his siblings, which Ying attributes to his being raised, as a baby, on milk, as opposed to rice milk, a mixture of ground rice and water. Elizabeth was already in medical school by the time Devon was old enough to have some of his earliest memories, which meant that one member of the family was already well on her way to joining the professional class; meanwhile, two other siblings were attending selective universities. The family's finances, should there be any worries, would be managed by his adult siblings, who could step in if a tax complication or financial glitch emerged at the restaurant. Although Devon worked hard at the restaurant, at school, and on his (various) instruments, he was unburdened by whatever serious troubles arose for his parents at the restaurant, as his siblings shielded him from them. The older siblings couldn't help but spoil him—Elizabeth loved treating him to fancy meals—even as they worried that he wouldn't have the same drive that they all did, because his early years entailed less responsibility.

By the time Devon attended the local high school, it had lost considerable funding, as the tax base suffered from a financial down-turn and the city's population was dwindling. Some of the opportunities that the older Chen siblings had enjoyed—a trip to England to perform along with other marching bands in a sprawling Christmas parade, for example—were no longer on offer. The high school's prestigious International Baccalaureate program, a challenging, accredited academic program for strong students—one that the first three Chens had all participated in—had also ended. When Gang went to All-State Band, the school commissioned a plaque that had room for three dozen more names. Every one of them has remained empty since he graduated.

It was true that Devon, like his siblings before him, was high school valedictorian. But he managed to secure that spot by too few points for his siblings' comfort; they worried that he watched too much television, played too many video games. "He did all right," said Elizabeth with a sigh.

When Devon was in high school, all three older siblings worried about how different his experience would be without any of them living at home to push him. Elizabeth hit on the idea of starting a book group from afar to try to make sure that Devon was at least reading challenging books thoroughly enough that he could keep up a spirited conversation with his siblings about them. Once a month, the three siblings would videoconference to discuss whatever they'd assigned him: They read *Sapiens*, a book of human history that President Obama had recommended and *The Best We Could Do*, an illustrated memoir of a Vietnamese immigrant's family story. They divided among themselves the responsibilities of helping him apply to college: One sibling worked with him on the essay, another on the applications, another on test prep. The people he always feared disappointing most, Devon said, were his older siblings—not his mother.

In the fall of 2018, Devon enrolled in Dartmouth. Of all the children, he'd been the least serious about his music training, even though he played the most instruments—piano, violin, and saxophone. He'd taken the path of least resistance in the face of his mother's demands: He went through the motions of practicing, he said, never really taking full ownership of his skill as a source of comfort or pleasure. Never sure when his mother would fly into a rage, he had a stomachache for much of his childhood. By high school, the physical blows had stopped, and he'd separated himself emotionally from his mother's demands that he practice more—he could see by then how irrational they were—but he kept going anyway, simply because it was easier than addressing her emotional volatility. Devon is grateful to the music teachers who helped him in Bristol; but only when he was living away from his childhood home, at Dartmouth, was he able to fully enjoy the classes in music history and the ensembles that he chose to pursue, to the surprise of his siblings. Devon, who ended up double majoring in music and computer science, had to relearn his relationship to music in a new environment in order to enjoy it—to play the saxophone for the music, not as some test of his self-worth or out of fealty to his mother's desires.

Devon graduated from Dartmouth in 2023 and landed a job at Amazon. He would get some professional expertise, earn a good living, and move to Seattle, where none of his siblings lived. He was taking a job at a behemoth, rather than following their own entrepreneurial examples—the example that, in fact, their own parents had set. "My siblings are live-to-work people," he said. "I'm more work-to-live."

As someone who works in tech, Gang said that he sees in his mother the same force of will, one that can even seem delusional, that he sees in some tech founders—that outsized conviction that they can move mountains, bend people to their will, overcome tremendous obstacles. Sometimes, he acknowledges, Ying even did: From the unceasing grind of their own work life, she and her husband catapulted all four of their children into worlds of great privilege. Those children can feel gratitude, even awe, at the sheer amount of work their parents took on in order to launch them, while also feeling powerful mixed emotions, even anger, as they look back. Gang knows the effects linger—in how excruciating he finds it at work, for example, to contemplate the possibility that he might have made some kind of a mistake.

Long after her children's professional careers were underway, Ying was still giving her children career feedback that none of them asked for: Shouldn't Devon go to graduate school? Maybe even Gang should think about graduate school, she sometimes suggested. Education was almost as good as real estate, something you could own, even if you couldn't touch it, something that could always help you grow.

For a long time, she still had dreams for her children that were really for her—dreams of the family reunited. So many parents want their children to conquer the world, without thinking through the consequences of how big the world will be, how far away their ambitions, once realized, might take them. In recent years, Ying has had the luxury of briefly closing the restaurant to visit her children; it's not quite the same, though, as having grandchildren close by, something a few of her relatives nearby can enjoy because their children went into the family restaurant business.

Having launched her children into glittering lives, in cities of opportunity and connections, she still sometimes thinks better of their decision to leave Bristol. San Francisco, where Gang now lives with Steph and their two children, strikes her as unsafe by comparison to Bristol, a city free of earthquakes and historic fires. All things considered, she sometimes tells Gang, she thinks he should think about moving home.

EIGHT

Finding Nurture

Augustine Kong, a professor of statistical genetics at the University of Oxford, and the author of a paradigm-shifting paper on genetics, also grew up in a Chinese family that saw its fortunes rise and fall and rise again. Augustine's parents moved to Hong Kong before the Communist regime took power, a move that cost them much of their once-considerable wealth. His parents never fully regained the financial status they had had, but his father managed to open a factory, earning enough and saving enough to send seven family members—Augustine, his four siblings, and his aunt and uncle—to schools in the United States in the 1970s. Augustine believes that his paternal grandfather, a highly educated man, was the force who drove his father to ensure those educational opportunities for the rest of the family. "My father probably would have felt he failed my grandfather if he did not do everything he could to support our education," said Augustine.

Augustine majored in math at Caltech, then went on to receive his PhD in statistics at Harvard before eventually stumbling into genetic research, publishing, in 2018, a paper in *Science* considered among the most important papers of that year in the field of genetics, a work that would inspire many scientists to reexamine their own data in a new light. It was work that could provide a new approach to analyzing the way that parenting can help children succeed—using genetics as a tool.

From just a small sample of collected saliva or blood, researchers can now regularly glean most of an individual's genetic building blocks, the new and the inherited mutations that contribute to how each person's development unfolds. In recent years, as DNA testing has become more sophisticated, statisticians and geneticists have been able to analyze data on the individual genomes of millions of people, in numbers high enough that geneticists can detect small but meaningful associations between numerous genetic variants and certain aptitudes and outcomes.

At first, researchers such as Augustine were primarily interested in establishing links between those patients' genomes and their health: What gene variations—and there were likely thousands of them—were associated with body-mass index, for example, or cancer risks?

Because educational attainment—how far someone goes in school—can be a factor that researchers would want to account for in their analyses, they often asked those individuals to provide that information as well. Over time, they ended up gathering enough genetic data on educational attainment that researchers could recognize hundreds of genetic variants associated with the length of time that individuals stayed in school. Since then, as more data has been gathered, the number has grown to upwards of four thousand. The associations are significant for research, but so subtle that they could not meaningfully predict results at the level of an individual.

Augustine worked at an institution in Iceland that had access to a large genome dataset that would allow researchers to compare about 21,000 Icelanders to their parents. Instead of looking to see how inherited genes affected children's lives, Augustine was curious to see what happened when those genes were *not* inherited. Geneticists would expect children who inherited genetic variations associated with college completion to be more likely to go on to finish college. But would that be true for those offspring who didn't inherit those genes as well?

That nontransmitted genes could nonetheless be hugely influential for offspring was a likelihood that animal researchers have long considered, a cause and effect known as "genetic nurture." For example, if a cow has genes that predispose it to make a lot of milk, it is likely to have calves who are bigger and healthier and can therefore make more milk, even if those fortunate calves did not inherit the specific genes associated with prodigious milk making. How genetic nurture applies to humans, however, has only recently become a subject of considerable interest.

What Augustine found was something that could confuse, delight, or dismay those who emphasize the influence of the environment and hereditarians alike, depending on how the findings are interpreted. Augustine found that the combined gene variants associated with educational attainment still had a significant effect on the offspring's educational attainment even if the offspring did not inherit that part of the genotype. (The effect was roughly 30 percent as great as the effect of genes that were, in fact, transmitted.)

Parents' genes seemed to have some influence on their children's educational attainment, even when they weren't inherited—which would mean that some researchers had all along been attributing too much power to the effect of genes alone and were underestimating some aspect of the nurture effect.

"I think Augustine's paper challenged everybody," said Dalton Conley, the Princeton sociologist, who also has a PhD in biology. "Because now geneticists are saying, 'Wait a minute—what you're calling a direct genetic effect is actually partially environmental.' But the geneticists can also say to those social scientists, 'You know what? Even when you're measuring the environment—the effect of parenting—what you're actually measuring is the metagenome. It's just genetic effects, one degree removed.'"

Educational attainment, as a measured life outcome, has always been somewhat unusual in the field of behavioral genetics. How much schooling a child gets is one outcome that researchers can see is strongly influenced by a child's upbringing. Fraternal twins may dramatically differ in their height, their divorce rates, or their level of extroversion—but they tend to pursue a college education at relatively similar rates (even if those rates aren't as close to each other as are those of identical twins).

So much of the power of parents is bound up with their influence on how much schooling their children obtain, especially now, especially in the United States. College graduates are more likely to marry, stay married, own homes, and enjoy psychological benefits associated with marriage than those with only high school diplomas. Most economics researchers believe that, even now, with college costs so high and the risk of debt a deterrent, a college degree yields financial payoffs in the long run.

But in the study of the genetic nurture of educational attainment, what, exactly, is the equivalent of the cow's copious milk? What is the expression of those genes—the parenting effect—that is tied to those outcomes, even in kids who don't inherit those genes? Augustine's paper did not try to answer the question of how those nontransmitted genes—how the parents' behaviors—were affecting educational outcomes. It would almost certainly be an elaborate suite of qualities, but Augustine feels strongly about one conjecture: Being a parent who went to college (as one example of an outcome) was not a requirement for creating the environment conducive to a child's going to college. Many of the mothers in Iceland who had many of the gene variants common in people who actually did achieve high rates of education had not, themselves, even attended college; few women did, in that cohort, when they were the appropriate age. Nonetheless, those genes worked on their children indirectly.

Since then, other geneticists have taken issue with possible environmental confounds in Augustine's work, without ruling out the possibility that his findings will still yield great insights. If the research holds up, said Daniel Benjamin, who has collaborated with Kong, "this would be a whole new approach to studying how parents matter." Researchers might try to discern how those genes influence parents—that is, what behaviors they are associated with—in ways that end up mattering for the children, whether they inherit those genes or not. This information might provide an innovative way to enrich the body of parenting research that exists to date.

Augustine is not a social psychologist or a developmental psychologist, but he was once someone's child and, like everyone else, reflects on the way his upbringing may have shaped him. His father's heroism in procuring an American education not just for Augustine but also for his four siblings and his aunt and uncle clearly moves him. "No one is perfect," he said, "and my father was by no means perfect, but he did do this one extraordinary thing." His life's greatest contribution, his son thought, was setting in motion so much possibility for seven future lives.

Science springs from theory, and theory often springs from personal experience. "I actually believe—at this point this is purely speculation—that a substantial amount of what is captured by strong educational attainment variants are not, so to speak, variants that make a person smarter or not," he said. "I believe they are what you can think of as planning variants, or what I referred to as delayed gratification variants. These people are more prone to plan for the future." They are people, Augustine theorized, who can endure present pain for future gains; people who tell their children, as Augustine's father often did, to think of their futures; people who will suffer through the drudgery of a job or a class for the rewards that arrive when that drudgery (and there is always drudgery) has been completed. They are people who might be willing to sit in a hotel room in Fuzhou alone for three days, weeping, thinking about the children left behind but also about the future that would lie ahead. They might be willing to alternate years of schooling in Tuskegee with years of farming, far from home, even at a young age, confident that the payoff of a college degree would be worth the wait.

They are probably also people, Augustine went on, who think about the future in a particular way. "For a person to have hope is very important," he said. "Because if you lose hope, you lose motivation. They have this optimism, or this hope, that if they do all of this, things will get better. And once you have that motivation, you put in the effort. But if the situation is bad and you accept that it will not get better, then you will not put in the effort—you'll have no plan for the future."

Augustine, the math major, the statistics PhD, the geneticist, seemed to have landed on some version of that classic principle of social psychology: expectancy-value theory. Without the expectation of a happy result—without at least hope—there would be no motivation to pursue whatever was deemed of value. It was almost hokey: the idea that people who just have a little faith will go much further, take the chance of enduring the present tedium—or loss, or pain, or suffering—for some future, wholly unguaranteed joy. But are those gratification delayers people who have blind faith, a more general optimism? Or do they simply have more faith in themselves?

ONCE, WHEN GANG CHEN WAS AT YALE, a fellow student, a Yale legacy, started a conversation with him about the eternally fascinating question of nature versus nurture. The fellow student was trying to explain to Gang the limits of one's genes—that even if he himself practiced basketball every day, ten hours a day, he would never be a truly great player. And maybe this friend had been reading some reactionary pop science, because he also suggested that genes were what kept certain populations from climbing out of the lower class.

"In that moment, I really did not know what to say," said Gang. He was self-conscious about outing himself as having come from what his friend would consider a lower-class household. He did not know what that student would think about Gang's parents. If the family had all stayed in China, and government policy had remained punitive toward families with three children, they would have almost certainly remained in the lower class indefinitely. That status would not have reflected anything about his family's abilities—it would instead have been the result of an inhumane government policy that clashed with the desire, imperative to the perpetuation of the species, to bear children. Gang knew that environment mattered and that cultural context mattered. He didn't think his friend's SAT scores were higher than his because his friend, as he suggested indirectly, had better genes; he thought it was likely that his friend's family had paid for all kinds of expensive test prep that was not available to Gang. But he couldn't begin to explain any of that to his friend, who knew only one kind of environment, the kind with financial cushions softening bumps in the road.

Even Gang, for all his innate talents, might not have ended up with the opportunities that came his way had he not had so supportive a community in Bristol, as well as a truly unusual mother. Ying was fiercely focused on attracting and drawing in highly educated people in the Bristol area to be active influences in her children's lives and found ways to offer her children some of the luxuries that families of the Chens' income level typically could not afford. Her children, when they were quite young, after moving on from their first, free piano lessons, eventually took lessons from a woman with a master's degree in music who taught at King University, a Christian college nearby. When Devon started taking piano lessons from two teachers at once (to Gang's disbelief), it was because the second teacher had a doctorate in musical arts from the University of Michigan, providing an opportunity that Ying simply could not pass up. Devon's violin teacher, early on, was a woman who was not just in the local symphony but was the concertmaster there.

The music lessons the children received translated into skill and opportunity—but maybe even if those instructors had not been teaching the Chens music but had simply spent quality time with them, one-on-one or in small groups in their company, they would have improved the Chens' prospects.

In 2008, a group of neuroscientists at the University of Oregon, led by Helen Neville, published a study that they originally thought might prove the special merits of music instruction in helping disadvantaged young children. The study, small but intriguing, was designed to test the hypothesis that music training improves "several diverse aspects of cognition," in part by improving attention. For the study, some students who were already enrolled in Head Start, a federal enrichment program for disadvantaged preschoolers, studied music every school day for less than an hour, in very small groups. But the researchers also set up groups in which students received, instead of music training, attention and focus training, with equally low student-teacher ratios. Yet another set of students were given typical Head Start curriculum instruction, also in very intimate groups, and others were given the same curriculum, but in a classroom with a typical size—about twenty children for every two teachers.

The researchers found that the students studying music did, in fact, see improvements in focusing attention, fluency with numbers, and solving certain cognitive puzzles. The same was true for the small group of kids who were getting instruction on their attention skills—but it was also true in the small group that was going through the usual Head Start curriculum, with no emphasis on music or attention. The only group that did not show improvement was the group that was in the large Head Start group.

One tentative conclusion of the study, the researchers wrote, was that the gains seen in children in small groups "may derive from the fact that music training typically involves time being individually tutored, or being in a small group, which may itself increase opportunities for training attention."

Parents may assume that musical training has some focus-enhancing powers or cognitively enriching benefits. But Neville's research suggested that music instruction might also be valuable because it's generally one of the rare extracurriculars in which students spend one-on-one time with a teacher—often a highly educated individual—while learning.

It wasn't just that Ying advocated on the Chen children's behalf. From the beginning, their community did, too, providing the kinds of opportunities that Neville's research suggests could be so valuable. When the Chen children were small, they had, compared to their American peers, copious amounts of time in the intimate company of educators. Early on, the school provided them, as a small group of three, with an English tutor at school. After school, they studied English, also only in a group of three, with Angela Thorp, who worked with them as an ESL volunteer; also highly educated, she had a graduate degree in aquatic ecology and would go on to work in the administration of the University of Wisconsin at Madison's computer science department. Angela continued to spend time with the children weekly long after their English skills were strong. Having grown close to the three of them, she tried to help them express their creativity. "I remember once we did a play for Elizabeth's birthday, based on a Chinese folktale, 'The Golden-Haired Princess,'" said Thorp. "They acted it out." She made up games with them, and then, when they were older, took them on outings—their first visits to a theater, their first meals at restaurants other than the Shanghai Chinese Restaurant. Eventually, as the three were clearly thriving in school, Thorp became more of a mentor and family friend.

By the time Devon was twelve years old, his three older siblings were too far away to help him in any kind of daily way. But his mother, over the years, had procured adult friends, influences, role models.

To Gang, when he came home, Devon's packed musical instruction schedule seemed excessive, as if his mother had gone off the deep end with a last-ditch frenzy of ambition for her last child. Ying said she was just trying to keep him away from video games, but she was doing something more—she was finding a way, in an increasingly resource-challenged part of the country, its school budget slashed, to provide her son with the company of educated, engaged mentors in a one-on-one setting.

Ying clearly was not in touch with how much of that instruction her son could take in, or how much it cost him in time he would otherwise have spent relaxing or socializing with friends. But in some ways, her instinct may have been sophisticated: A robust body of research over the past decade has found that regular, high-quality tutoring improves students' academic achievement—more so than various interventions such as decreasing classroom size or supplementing with summer school. "Individualized instruction is among the most effective education interventions ever subjected to rigorous evaluation," wrote Matthew Kraft, an associate professor of education and economics at Brown University, in a 2021 paper. Subsequent research, which Kraft detailed in a 2024 working paper, found, in analyzing some 265 trials, that across the sample, tutoring improved students' academic achievement by the equivalent of about a year of learning in middle school. The study also found tutoring effects decline as the size of the program increases—which Kraft sees as a problem to be solved, rather than a refutation of tutoring's great promise.

Some of the best research on the importance of individualized mentoring has already happened in Germany, where a group of college students started a program called Rock Your Life, sending hundreds of students into the homes of disadvantaged families (one-quarter of whom were immigrant families) at least once a week. The results were significant for adolescents, substantially improving economists' measures of their value to the labor market. And it worked with smaller children, even achieving the seemingly elusive goal of influencing personality—the children became more "prosocial" than they were at the outset; they showed more signs of kindness, of defending others, of helping, cooperating, comforting.

The burden of having children become their more curious, engaged, academically challenged selves should not, in the end, fall entirely to parents, especially not parents who are struggling financially, like the Chens. Even if parents strapped for money have the temperament and the talent to coach their children carefully, they are likely short on time; it is time-consuming to make ends meet, to need two jobs or to work long hours at one, to go to three different stores with three different sets of coupons, to fix whatever is broken instead of hiring someone else to do it. Ying's children benefited from her exceptional resourcefulness and charm; the same force of will that caused much distress in her children also allowed her to offer them extraordinary opportunities.

Augustine would say that Ying's most influential quality, the one from which her children would benefit the most, was in her ability to take a long view when she made decisions for her family—the sacrifice of those eighteen months apart from her children, the long hours at the restaurant. She gave them what she could, which was too much some days, not enough others, accomplishing, somehow, her wildest hopes for her children, a feat they recognize. Elizabeth, now a mother herself, sounds like someone who has made her peace with the way Ying raised her. "She parented us how she knew best," she said.

NINE

The Paulus Family

I n addition to conducting casual observational research on other
families as a child, I also loved reading about them. I was especially obsessed with the rites and rituals of *Cheaper by the Dozen,*
the famous memoir by two of the twelve children of Frank Bunker
Gilbreth, the industrial efficiency expert. I longed to be expected, as
the Gilbreth children were, to fill out process and work charts that
would account for, among other things, my daily height and weight,
which I'd dutifully plot on a chart, after completing another chart that
indicated that I had bathed and brushed my teeth. I liked to think I
would be the fastest to arrive whenever Gilbreth blew his signature
summoning whistle. I relished the idea of all that focused rigor; I worried that I had potential for discipline that might go untapped. I expected to find, in some of the unusual families I explored for this
book, extreme trainings such as those practiced in the Gilbreth or
Kennedy house, or depicted in the movie *The Royal Tenenbaums,* in
which a widower played by Ben Stiller races his two young sons
through a high-stress drill to practice escaping their own home in the
event of some unspecified emergency.

Instead, I sometimes found that the parents in those families, such as Alfredo and Amalia, were strong, respected members of their community who were too busy keeping their family fed and clothed to indulge in the luxury of what I think of as "enterprise parenting"—parenting explicitly designed to build useful skills or stoke ambition or unify the siblings around some future goal. Sometimes, the parents I learned about in these unusual families were even all but absent by necessity. At one point early on in my reporting, I was introduced to an immigrant family in which all four children had grown up to become tenured professors, but they ultimately decided not to share the story of their upbringing—it was too painful for them to talk about the extent of neglect they had endured while their struggling parents tried to scrape together a living. Whatever made their childhood unusual, it was the opposite of enterprise parenting.

We also tend to hear only about the enterprising techniques of parents whose children's unusual success seems to be proof of those parents' efforts; we don't hear from the adults whose parents had them reciting can-do mantras, but who nonetheless found, over time, that they could not—could not finish the novel or make the national team or revolutionize the tech industry.

Instead, we are offered illustrative reminiscences from high achievers such as Denise Morrison, who posted on LinkedIn in 2013 a short essay called "Life Lessons from My Dad." At the time, Morrison was the CEO of Campbell Soup, and her sister, Maggie Wilderotter, was the CEO of Frontier Communications (she is now chairwoman of Docusign). Denise and Maggie, who have two other sisters who were executives in corporate America, were the first two sisters to be running big U.S. public companies, *The Wall Street Journal* reported in 2011. In her post on LinkedIn—and in a half dozen other interviews that either she or her sister have given—Morrison captures a high-achieving family culture she credits to her father, Dennis Sullivan, a former AT&T executive and classic enterprising parent. In the Sullivan household, the children received allowance bonuses when they

got an A, but saw that disbursement docked when they got a C or
below, as if school were their job and they were highly incentivized
employees. Each sister had to read a book a week and report on it,
orally or in writing, as she preferred; each week a chore, written on a
slip of paper, appeared in the sisters' respective job jars, a household
contribution to be completed by the end of the week (or swapped with
a sister's, an incentive to negotiate duties that might well have been
inspired by a similar arrangement in *Cheaper by the Dozen*). Their father roused them at 6 A.M., so they could do calisthenics as a family,
reported a *Wall Street Journal* article in 2007. "In my household, you
could not say, 'I can't,'" recalled Denise in the book *Earning It: Hard-
Won Lessons from Trailblazing Women at the Top of the Business World*.
"My parents would say, 'That's not in the dictionary.'"

As engaging as they are, the stories of enterprise parenting—which
make apparent how much discipline and focus such an approach entails on the part of the grown-ups in charge—can also feel exhausting
to contemplate. It's for that very reason that I have a particular fondness for the ways of an enterprising parent named Laurence Paulus,
whose unusual approach to enrichment was less about achievement
and more about a desire to make sure his children did not miss out on
his own passions.

Laurence, a producer of fine arts television programming, had a
modestly successful career at WCBS-TV in New York. He and his
wife, Teruko, raised their family of three children in a cramped two-
bedroom apartment in Manhattan. Laurence's musical education
was self-directed and spotty; he never made art of note. And yet he
was one of New York's great hidden characters, the kind of person
who thrives on the city's energy and, in his enthusiasm, generates
more of it.

Laurence often swept his family up in his zeal, as he did on the
evening of September 16, 1966, when he and Teruko headed to West
Sixty-sixth Street with their children: Janet, an eleven-year-old; Stephen, a ten-year-old; and their third child, Diane, born just two

months earlier. The family parked themselves near the fountain at Lincoln Center, as hundreds of "Great Society overachievers," many dressed in tiaras, white tails, and chiffon, *The New York Times* reported, walked a red carpet toward a new building's gleaming bright windows. They were there for the opening night of the Metropolitan Opera House, which was showcasing a never-before-seen production: Samuel Barber's *Antony and Cleopatra.*

Inside the theater, the performance was received with some disappointment, was even regarded as a fiasco ("Almost everything about the evening, artistically speaking, failed in total impact," declared *The New York Times* review the next day). Outside, Laurence nonetheless felt close to greatness. The family could not afford tickets, but he wanted to be there just to feel a physical proximity to that momentous event in cultural history. He had the foresight to bring a transistor radio, so the family could sit outside the new building and enjoy a live broadcast of the event. The radio was in Diane's stroller, and, as a result, a crowd of other opera enthusiasts gathered around the family to listen. Laurence often recalled this family experience around the dinner table, proud of being something of a (peripheral) star of this quintessentially New York tableau. The indelible image that lingers is of Diane, an infant, perhaps sensing the drama and electricity of the moment, the attention of rapt strangers around her—a feeling of being at the center of some mysterious force that she would eventually spend her life trying to harness and share with others.

Paulus and his wife, Teruko, who was born in Japan, were, in many ways, ordinary people but unusual parents, the kind who saw child-rearing as not just their life's work but their highest form of creative expression. Their oldest, Janet Paulus, grew up to be a professional harpist who has performed with some of the world's leading orchestras. Stephen Paulus, the middle child and their only son, played a crucial role in building NY1, an innovative local New York news channel that served as a model for many others. Their youngest, Diane Paulus, is considered one of the most influential contemporary theater directors of her time, an innovator who expanded mainstream audiences' idea of what theater could be.

"I think about my parents all the time," Diane Paulus told me. "One of my mentors, Andrei Serban, a very famous Romanian theater and opera director, used to say, 'You know a great actor onstage because they carry not just themselves but their ancestors.' And it's such a beautiful thought. But it's really true—it's like they are, in that moment, not just themselves but everything that's come before them. I really think about my parents that way."

LAURENCE PAULUS, the son of a German immigrant firefighter, first developed a love of theater in the early 1930s, when he performed in local summer stock as a young man and discovered that he had a talent for the stage. He considered pursuing a career as an actor, but instead, in 1941, he was drafted. Theater, as the family story goes, may have spared his life: In an interview for a post as an army clerical worker, he quoted a line from William Saroyan's play, *The Time of Your Life*—"No foundation. All the way down the line." Although context is lost to history, the reference apparently impressed his interviewer enough to land him a desk job (it also helped that he could type). Laurence never saw active combat, remaining stateside for the duration of the war, working his way up to the position of first lieutenant.

He went overseas only when the war was over, opting, when given the choice, to spend his last days in the military in Tokyo, as part of the Allied forces' occupation of Japan. There, he seized theatrical opportunities that he could only have dreamed of in the United States. He started directing crowd-pleasing plays for what was known as the "Radio City Music Hall of the East," the Ernie Pyle Theatre, the newly renamed Japanese theater seized by the American occupying forces. He mounted shows for audiences of three thousand people; he acted in a production of Noël Coward's *Private Lives*, playing Victor, an upright, spurned husband, part of an ensemble that landed "great gusts of laughter" from the packed audience of "officers and high-ranking citizens from the various occupying forces," wrote a reporter for *Stars and Stripes* in his review.

In 1947, Laurence, then thirty-three, was shopping at a department store when he met a young Japanese woman, then nineteen, who was working behind a counter. He struck up a conversation when he noticed that she had somehow cut her finger. "Ouchie ouchie," he said to her, pointing to the finger, assuming her English was poor. "Yes, I cut my hand," Teruko said. The story was emblematic of so much about what would unfold in their relationship, a partnership built early on Laurence's paternalistic if well-intended impulses, with Teruko, a fiercely competent woman, accepting that, sometimes with forbearance.

Laurence was gregarious even by American standards, a member of the conquering forces; he arrived confident and upwardly mobile, indulging a passion for theater he had always dreamed of pursuing professionally.

Teruko, by contrast, had seen the world as she knew it crumble. The college she'd been attending had burned to the ground. Her father's import-export flower business had dissolved shortly after the attack at Pearl Harbor. During the bombing of Tokyo, she had slept in cemeteries, hoping she would be spared and safe. For the rest of her life, she avoided fireworks, the sound too much like the blasts she had heard when she walked home from school as bombs exploded a block away. By the end of the war, she was an orphan. As a seventeen-year-old girl, she transported her brother, emaciated from his service in the war, to safety, carrying him on her back for miles until she reached a camp where she could nurse him back to health. Another older brother died serving in the military of Japan.

Beyond the differences in their circumstances, Laurence and Teruko were also dramatically divergent in temperament—Teruko, disciplined and focused, Laurence, jocular, a showman at heart. Even so, their children can imagine what drew them together when they met: attraction, a shared love of the arts, and a longing for a sense of grounding. Defying anti-fraternization regulations, the couple found a small hideaway in the hills of Tokyo where they thought their relationship would go undetected, an idyll that ended when their home was raided by the military police. Almost all their possessions were seized. Teruko was briefly detained; Laurence was suspended from his work for a week. Their relationship was defined by secrecy and uncertainty, the plot of their story practically operatic. And then, after all they had been through together, in 1948, when Teruko was twenty, the romance ended just as dramatically: Laurence left Tokyo, eager to travel the world.

"I've only been sailing for four hours but I know that it's going to be strange—and hard—not to see Teruko each day," he wrote to friends two days after Christmas in 1948. "She is a swell girl and really deserves better than I have now left her to face. It will be a real test of what she is made of. I often wish I could look into the book of time and find out if she would have been my destiny. However, after being an independent bachelor for thirty-five years, it isn't easy to think about settling down." And so—in a chapter that the Paulus siblings didn't fully understand until their father's letters emerged when they were already adults—Teruko joined the ranks of other women in her position, known as "poor butterflies," Japanese women who were abandoned by American servicemen. Laurence knew, he wrote home, that he could not hope for a more devoted woman; at the same time, he was not prepared to face a life filled with the prospect of "petty people with their gigantic prejudices," either in Japan or in the United States.

Laurence spent eighteen months traveling around the world—he started in Shanghai, then made his way to Bombay, Cairo, Marseilles, and London before he returned to the States, all the while remaining in touch with Teruko by mail and phone. Teruko started working with an older sister who lived in Los Angeles to try to gain entry to the United States, finally succeeding in arriving in Chicago, where another sister lived, in September 1950. With financial help from Laurence, whose career in television was just starting, she moved to New York to attend a small Catholic college on Staten Island and, at his request, converted to Catholicism. Just five years after the end of the war, Teruko, living far from one sister in Los Angeles and another in Chicago, chose to trust an American man fourteen years her senior who had already abandoned her once before. Her sisters disapproved of her possible union with an American, so much so that no one from her family would attend their wedding when they did eventually marry. Teruko was a private woman who rarely showed strong emotion; only toward the end of her life did her children start to grasp how much she still thought about the costs of that bet that she made on her future.

The two married in 1954, when Teruko—Terry, as American friends started to call her—was twenty-six and Laurence was forty-one. They had two children in quick succession, first Janet in 1955, then Stephen just under a year later, at which point they were living in Merrick, Long Island, part of the wave of GI families seeking roomy, affordable homes. In 1960, they returned to New York City, moving to the Upper West Side. Laurence was working at CBS in New York at an opportune moment for live television, and he wanted to be able to walk home from his office in time to see his children before they fell asleep. Within a few years, many families would be fleeing to the suburbs, but Teruko and Laurence stayed committed to their lives in Manhattan. The choice made sense for a couple so devoted to the arts and also their children: Laurence was the kind of deeply involved father one associates more with contemporary parenting culture—doting, focused on enrichment, prioritizing time with his children.

By then Laurence had moved on from any hopes of acting or directing, seeking cultural fulfillment instead in his job as a production manager for some of CBS's most lavish arts programming. He and Teruko made sure their two children took advantage of offerings available to them through his work. They attended the New York Philharmonic to watch Leonard Bernstein's beloved Young People's

Concerts, which aired on CBS. Laurence often brought Janet and Stephen to the set of *Camera Three,* a CBS arts show he produced that featured presentations on Salvador Dalí, dramatized the work of Fyodor Dostoevsky, and broke ground by casting a Black actor in a production of *Othello.* On the set of that show, Laurence introduced his children to the pianist Glenn Gould and luminaries of the Bolshoi Ballet; he brought home for dinner the conductor Alfredo Antonini, who directed CBS's thirty-piece orchestra.

And when Laurence decided that eight-year-old Janet should play the harp—it would make her unique, he thought, something he valued more than she did—he asked the harpist for the CBS TV Studio Orchestra, Pearl Chertok, if Janet could sit by her side during interludes when the orchestra was playing but she was not. Chertok, who agreed, was also a neighbor, they discovered. By then, the Pauluses had moved to Lincoln Towers, just a few blocks from Lincoln Center. The housing complex was a neighborhood unto itself, packed with artists, academics, writers, and musicians. Chertok—who became a well-known music educator—loaned Janet a small harp to practice on and never charged the family for lessons, out of affection for either Laurence or the young, earnest girl who paid such close attention.

That Laurence was ambitious for his children is clear from one of the two dozen or so favorite anecdotes that he told and retold around the dinner table. In 1962, the Greek Canadian soprano Teresa Stratas appeared on the CBS game show *To Tell the Truth.* Laurence spotted Stratas's mother in the audience and struck up a conversation with her, in part so he could ask her what he really wanted to know: How did she do it? What was the secret to raising a daughter who achieved such success? Her answer was the punch line to his story: "Lots of peanut butter!"

That he liked to tell the story, with its absurdist takeaway, suggests that Laurence had some self-awareness about the line of questioning that prompted her comedic response. All three of the Paulus children agree that their parents were determined that they participate in the arts; and yet they also agree that they never experienced their parents' clear mission as unwanted pressure. Laurence had a boisterous, contagious enthusiasm; Teruko was supportive and uncritical. If they had still been living in Merrick, Laurence and Teruko might have encouraged their children to audition for community theater. But this was Manhattan, and the local theater, for them, was Lincoln Center. In 1964, when the New York State Theater (now the David H. Koch Theater) was preparing to open its inaugural production, *The King and I*, Teruko brought Janet and Stephen, then nine and eight years old, respectively, to audition for chorus parts as the King of Siam's children. Stephen recalled a crowded room full of white children wearing kimonos, with their eyes heavily lined to make them appear more Asian; his mother made no such effort with her own children. The two children were asked to crouch, jump, and run, which they did as best they could, but to no avail—neither was offered a part, or at least that was their understanding. Stephen only learned as an adult that he had, in fact, been offered a role in the production, but Janet had not. It was both or nothing, Teruko had clearly decided. The purpose of the arts was to provide pleasure, to build confidence; there was no value in pursuing an opportunity for Stephen if it would be experienced as a painful rejection for Janet.

It was a near miss, but the family appreciated that an audition for a production of that caliber was even on offer; Manhattan, for Laurence, was an ideal setting for those kinds of close encounters with cultural events and heroes, all of which became part of his personal mythology as a New Yorker. "Bum bum bum BUUUUM!" Laurence once bellowed to the conductor Zubin Mehta, singing the famous first notes of Beethoven's Fifth, when they passed in the street. ("Wrong key!" Mehta called back.) Diane was told that when she was a small child, she and her parents crossed paths with John Lennon and Yoko Ono, who had not yet had a child together. The couple stopped in their tracks, Laurence told her, at the sight of the father and daughter. The Paulus parents were not famous, but as an interracial couple with children as early as the late fifties, they were trail-blazers.

The streets of New York were an always-available stage for Laurence, which could be excruciating for Janet, who was self-conscious as a child. She knew, if he grabbed her arm as they were walking down the street, that something bad was going to happen—and that thing was usually that he would burst into song at the top of his lungs. "He knew I was very shy," she said. "And I just wanted to go hide somewhere—but he grabbed me, and I couldn't get away. It was anything to get me to be less shy—to be more like a showman."

His impresario instincts, such as his plan to make Shakespearean actors of Janet and Stephen, were better received when they happened in the privacy of their own home. In the early 1960s, the first years that the director Joseph Papp brought Shakespeare to the Park to audiences at the then-new Delacorte Theater, Laurence helped WCBS-TV, the local channel, live-telecast the plays. Janet and Stephen attended a 1964 Shakespeare in the Park production of *Hamlet* (which ran on the local channel for three hours without commercial break). Laurence wanted to see his own children performing in those Central Park productions one day—why not Stephen as Romeo or Janet as Juliet? Janet was about ten when her father had her and her brother memorize the lines to the balcony scene.

"Romeo, Romeo, wherefore art thou, Romeo?" Janet recited when we spoke, offering proof that she remembered the lines even now. "Deny thy father and refuse thy name. Or if thou wilt not, be but sworn my love. And I'll no longer be a Capulet."

Exhibitionism, in general, made Janet uncomfortable, but she appreciated her father's investment in her cultural enrichment. Before the family moved to Lincoln Towers, they had lived farther uptown in a smaller apartment, which had a map of the world on a wall. During dinner, her father would use a long pointer to point to a country on the map. First, he would ask Janet to name the country; then he would ask her to name five cities in that country. "It wasn't like he was testing us," said Janet. "It wasn't forced down our throats. It was more like constant learning. It was just what we did."

When Janet was starting high school, her parents enrolled her in the National Academy of Ballet, a small school founded by the Russian instructor Thalia Mara. At home, her parents were even-keeled and supportive, but they must have been drawn to the highly disciplined mode of instruction offered at the school. Janet was tall for her age and stooped, which is one reason her parents chose a school of ballet; but in class, she was often made to feel that her height also made her unsuitable for ballet. The demands of the school were sometimes harsh: If a student was seen with hair slipping from her bun, she was thrown out of class. As a fourteen-year-old, Janet went home every night to iron the ribbons she affixed to her ballet slippers. "We were instilled with an education based on hard work, respect, humility, and honesty," she said of her time at the school. "So even at times when the truth hurts, it's something we accepted openly." It was as if the Paulus parents outsourced the cultivation of Janet's discipline, but made it bearable with a homelife that was calm and warm. Were both environments—or were both parents, for that matter—hard-driving, Janet might have rebelled or quit.

Teruko's own discipline and work ethic was always on display in their small apartment, which doubled as her workplace. A skilled seamstress, she started a small business out of their home so the couple could afford to send their children to private school. Often sewing late into the night in a hallway outside her bedroom that was crowded with bolts of fabric and other supplies, she developed that workshop into a one-woman thriving home design enterprise, making drapes, throw pillows, elaborately scalloped place settings and tablecloths that she marketed and sold to Park Avenue boutiques. Eventually, she had a thriving clientele of wealthy individuals in Palm Beach as well. Jacqueline Kennedy was pictured in *Women's Wear Daily* wearing one of Teruko's tablecloths, which had been fashioned into a skirt, an image so widely admired that Wallis Simpson, the Duchess of Windsor, asked Teruko to her home at the Waldorf Towers to help her achieve the same look. Teruko was almost always home in the family apartment in Lincoln Towers, but she was also almost always working there, filling orders, earning enough that she and Laurence could offer their children the opportunities they wanted to give them in Manhattan—not just private schools, but classes, theater tickets, costly musical instruments.

The projects that Teruko took on were creative, but there were always more orders to fill. Diane recalls a Broadway show she was thrilled to attend, at age eleven, with her mother, only to be disappointed when Teruko fell asleep soon after the lights dimmed. "I remember feeling so upset," Diane said. "Like, how could she fall asleep during that?" As an adult, she said, her heart breaks when she considers how exhausted her mother must have been—how rare it was for her to stop and sit down.

If they didn't have the money for some form of artistic enrichment they wanted for their children, Teruko or Laurence relied on their own talents to provide it. In order to afford a piano, Laurence learned from a friend, a well-respected piano tuner and instructor, how to take a keyboard apart himself and put it back together in perfect form. He frequently bought old pianos at auction, fixed them up, and sold them when another, better instrument came on the market that he could build back to its full musical potential. The first harp Janet owned was an old Wurlitzer Starke model, purchased for $170—even then, in good condition, they sold for more than $5,000—which Laurence managed to restring with the help of another friend who was a harpist.

As a girl, Janet spent her summers at a prestigious music camp in Michigan, studying harp. When the school offered her a partial scholarship to attend its boarding school year-round, starting her sophomore year in high school, her parents were thrilled. She accepted the partial scholarship from the school, now known as Interlochen Arts Academy, an important opportunity, Janet said, because she would both play in an orchestra and gain chamber music experience.

Janet's leaving home for a boarding school seemed to inspire Laurence: Now that Janet was studying on a partial scholarship, and because of the success of Teruko's business, they could afford to send Stephen to private school. Why not send him abroad? He could go to school in Switzerland, learn to ski and speak fluent French. In the summer of 1970, just before Stephen turned fourteen, he and his father traveled around Switzerland, visiting boarding schools and choosing one that Stephen ultimately attended, sometimes not seeing his family for ten months at a stretch, at a time when phone calls home were prohibitively expensive.

"It was my father's impetus," Stephen said. "He pushed us into things, and we said, 'Okay.'" Whatever Teruko's feelings about having two children far from home—Janet thinks it pained her deeply—she went along with it. "She would never express her frustrations," Janet said. "She put up with him—his craziness and his ideas."

Just as the ballet school provided Janet with a level of rigor that would be difficult for most parents to enforce, the boarding school was highly structured, especially for 1970: The boys received haircuts once a month, their relationships with teachers were formal, and lights were out at nine-thirty.

Even with Stephen at boarding school in Europe, Laurence kept considering other, even more far-flung adventures for his children. Wanderlust was practically a family value, one that Laurence reinforced with his stories from the eighteen months he'd spent traveling around the world: how he joined a puppeteering troupe in India, dined with the captain on a freighter to Venice, illegally transported an accordion from France to Italy, sampled the bathhouses of Athens, and cadged free three-star meals across Europe by claiming to be a restaurant critic.

Flipping through *The New York Times* one day, he read about a school at sea for high school juniors and seniors. The students would spend seven months on a Norwegian windjammer that would stop in North Africa, South America, and parts of the Caribbean. He asked

Stephen if he was interested—and Stephen said he was, ultimately leaving his boarding school for the school at sea. In the middle of the ocean, Stephen watched the northern lights flare through the night sky; he slept in a hammock, worked long hours as part of the crew, spent some days on the boat soaking wet and other days in ports in corners of the world few people he knew had visited. "It was the greatest experience of my life," he said.

As an adult, Janet looked back on the decision that she would attend the music academy with more mixed feelings, even some poignant regret mixed in with her recognition that the school launched her career. Her gratitude to her parents is bound up with awareness that taking the opportunity also entailed a loss—the curtailing of a childhood in her parents' home. "I never questioned anything that my parents decided, I trusted them implicitly," she said. "The cost— I didn't realize until later."

The way parents who love baseball might be quick to sign their child up for Little League, the Paulus parents looked for points of entry for their children into the arts. When Teruko sewed, she frequently listened to opera at high volume; opera, she told her children with great conviction, was the highest of all art forms. She enrolled Stephen, when he turned twelve, in a classical singing class, which he endured for all of two lessons before he quit. No one urged him to go back; and yet whatever successes Stephen went on to have, Teruko seemed to feel he'd missed his true calling as an opera singer, said Randy Weiner, an entrepreneur and fellow theater director who became Diane's husband, and who knew the family since he was in high school. Teruko voiced that regret aloud so many times Randy came to find it comedic, since there was no reason to think Stephen had any vocal talent (Stephen assured me he did not). "In forty years of knowing him, I literally have not heard him sing once," said Randy.

Laurence's two older children were different from him in ways that surprised and even irked him; his tutelage and modeling were not enough to make them more outgoing. When her father burst into song in public, Janet was desperate to distance herself from him, which is why he gripped her arm. "My father was very much frustrated with Stephen and me, because we were shy," said Janet. "It was like pulling teeth to get us to play the piano when people would come over. And then I started harp lessons, and he'd want me to play the harp for people, and I didn't want to do it. I was embarrassed."

By virtue of timing, Diane, who was born a decade after Stephen, was the family showstopper—unexpected, exciting, a novelty, a baby at the center of a rhapsodic audience at Lincoln Center that night at the Metropolitan Opera, the much-adored pet of her older siblings. Of the three children, her interests and instincts most naturally gratified Laurence's showman sensibilities. As a young girl, she would pose for silly photos, take her father's direction, put on whatever costume he proposed. "When Diane came along, he was in his glory," Janet said. "It was almost a relief."

Janet has a vivid memory of a British guest coming to dinner, a teacher from the boarding school Stephen attended in Switzerland. Janet would not play the harp for him, as she knew her father wished she would—but Diane, not yet five, stepped forward to pluck out the strings on the instrument, playing "London Bridge Is Falling Down" to an admiring audience. One can imagine Janet as a teenager, half-relieved, half-amazed by her sister's ease in fulfilling her father's wishes, in a way that she could not, despite the remarkable technical proficiency she'd acquired after all those years of hard work.

Laurence and Teruko were fifty-two and thirty-eight, respectively, when Diane was born in 1966. By then, the Paulus parents were marginally more affluent; they knew the workings of New York better; they had seen their children through various arts programs, compared notes with other parents. Still, the Paulus family was so far outside the circles that entered the most elite private schools, they could not get an interview for Diane, a problem that resolved when Diane's standardized test scores came back exceptionally high. Diane was enrolled at Brearley, an elite girls' school where she'd spend thirteen years of her education.

Laurence also applied himself to reinforcing whatever talents Diane seemed to possess already. When Diane was in grade school, he had her read out loud from *The New York Times,* correcting her pronunciation. He could put up with her watching television, but only on the condition that she used that same time to learn the skill that spared him from combat: typing. She was the rare ten-year-old girl who, while watching *The Bionic Woman* or a program on PBS, was simultaneously practicing speed-typing.

On a wall in view of the dining-room table, Laurence had created a kind of poster composed of postcards of portraits of classical composers, arranged in chronological order by birth. There, in the upper left-hand corner, was Handel with a long white wig, then Bach in his own shorter wig, then, moving to the right, Haydn, soft-faced Mozart, and others, going all the way up to Shostakovich, stern in his round black glasses. None of the images were identified by name, yet Diane, as a child, could easily recite, in chronological order, the name of each one, trained by a father intent on instilling in his daughter an appreciation for the greats.

Around the same time she enrolled at Brearley, Diane started taking dance classes at the local YMCA, as so many little girls do, but she quickly graduated to another local dance class: one held at the School of American Ballet, the "associate school" of the New York City Ballet, just blocks from her home in Lincoln Towers. Soon after that, she auditioned and was selected to dance with the New York City Ballet. At eight, she played the part of a child at the Christmas party in *The Nutcracker;* at nine, she danced under the eye of George Balanchine in a production of *Firebird;* at twelve, she danced alongside Baryshnikov in a production of *Harlequinade,* one of just four dancers on stage at the time.

A Very Young Dancer, a 1976 book of photojournalism by Jill

Krementz, about a girl who was in the New York City Ballet at the same time as Diane, captures the magic of seeing that heightened world through a child's eyes, as well as the stresses—the instructor who pressed a ten-year-old girl to hold in her stomach and who headed off anticipated complaints from mothers about casting choices. Diane came to think that the discipline, as challenging as it was, prepared her for the kind of harsh feedback that every person who enters the theater is exposed to, time and time again. "Certain students collapse under that," Diane told Columbia's alumni magazine in 2012. "I kind of thrived on it. To be told you're not good enough is amazing training."

Laurence was forever clipping news articles about possible creative venues for all his children, which is how he brought to Diane the idea that she might join the First All Children's Theatre, a Manhattan-based company founded in 1969. Its membership included only actors younger than sixteen, even as the theater launched original and "very serious" off-off-Broadway productions, as its founder and artistic director told *The New York Times* in a 1979 article. In judging whom they selected, the directors were looking for the following, according to that article: "potential talent, commitment to a tough and disciplined routine and the approval of parents, although the initiative must be shown by the child and not by an ambitious mama or papa." Diane, who joined the troupe when she was around ten, starred three years later alongside Ben Stiller in 1979 in a version of "Jack and the Beanstalk" called *Clever Jack*.

When she turned thirteen, Diane's parents felt like they had to make a decision. To help them make it, they turned to Janet, then twenty-three, to ask her opinion about Diane's future. Even as Diane was dancing and acting, she was achieving such proficiency in piano, which she practiced seriously every day after school, that they knew she was a strong contender to study at Interlochen. But should she? They put the question to Janet, who had by then graduated from the Cleveland Music Institute. Should they send Diane to live and study at a conservatory of some kind or, alternatively, enroll her in the kind of boarding school that Stephen attended, so Diane could perfect her French? Janet was relieved to have been asked, because she felt strongly about the answer. "No," she told them. "Keep her home."

Janet would have liked more time with her parents, not to mention with her baby sister; it was too late to change the course of her own life, but she could make sure Diane did not have the same regrets she did. Janet felt that no one in the family appreciated, at the time, how momentous it was for both her and Stephen to leave home to go to school while still in their early teens. Even when children seem to be highly motivated to maximize their potential in whatever they consider their passion, their parents must consider a question with no easy answer: Are those children old enough to understand fully what they might be giving up in order to pursue it? "When you leave, you don't really come back," Janet said. Her conviction was strong enough that her parents followed her advice. Janet often brought up her role at that pivotal moment in Diane's life in family discussions, a powerful reminder of how much she felt she had missed and of how responsible she felt for the more sheltered high school experience Diane enjoyed with the undivided attention of her parents in New York.

"Diane was so lucky," said Randy. "But Janet was the one who made it happen."

So much of that luck, Diane would say, was in the extra time Diane was able to spend with her mother.

Laurence's interactions with his children were playful and high-energy; but in the hours Diane spent alone with Teruko, daily, she received a quality of attention that she experienced as both steady and undemanding. She recalls being a young girl in first or second grade, directing her mother daily after school in an hours-long game of teacher, with her mother playing the student. Sometimes her mother stole time away from her work so that they could go to a park near Lincoln Towers, Damrosch Park, where Diane would dance and sing on an enormous bandshell where legends like Diana Ross would perform for crowds of thousands. Diane recalls keenly feeling her mother as an appreciative audience in those moments—she was offering the same quality of attention she did when Diane actually did dance on a grand stage with the New York City Ballet. Teruko attended every performance, even when it meant standing in the back of the theater for the duration of the show many nights in a row. She said little to her daughter about how she had danced on any given night; her presence was constant but benign.

"She had a deep way of looking without desire," said Diane. "It's the opposite of a parent who has desire—who is a stage mom and is pushing and guiding. And there are so many tropes of that for kids who become star athletes, gymnasts, whatever. Their mothers are pushing them. My mother came from a very Zen Buddhist Asian way of looking without desire—but was always *there*. Because you can't be seen unless someone is looking. I think she facilitated that for me my entire life, which gave me this inner sense of confidence—an inner sense of self."

For college, Diane attended Harvard, graduating in 1988 with high honors but without a clear picture of how she would begin to pursue a life in the arts.

She could see, from Janet's example, two important facts: that the life of the artist was possible and that the life of the artist was difficult. Laurence had encouraged Janet to become a harpist because it was unusual; but every orchestra or philharmonic needed only one. Simply moving her instrument required access to a massive vehicle of one kind or another. Janet reached some of the greatest heights of the profession—she played *La Bohème* and the *Ring* cycle at the Metropolitan Opera—but she and her husband, a cellist, also struggled to make a living for many years, before Janet eventually landed a position as principal harpist at the Orquesta Filarmónica de la UNAM in Mexico City, considered one of Mexico's premiere orchestras. Janet's life made it possible for Diane to understand what being an artist meant—and also may have given her permission to face its potential hardships without opting for a safer route.

Directly out of college, Diane started taking classes at the New Actors Workshop, founded by Mike Nichols, the director; George Morrison, a renowned instructor of method acting; and Paul Sills, cofounder of The Second City. Diane and Randy began pulling together quick and improvised productions in untraditional venues, producing *Twelfth Night,* for example, in a community garden on the Upper West Side. Eager to experiment in a less daunting environment than New York, Diane and Randy relocated to Door County, Wisconsin, at the suggestion of Paul Sills, who had a farm there; others they knew from the New Actors Workshop joined them. Along the way, Diane realized that the role of the director suited her skills and temperament better than that of the actor (an insight that clicked into place when her agent suggested she retake her headshots, so she did not look so much like "a Vietnamese bar girl"). In Door County, Diane and Randy conjured productions that called on all their theatrical training and overlaid it with the loose energy of artists who had little to lose and plenty of room to experiment. They engaged a local bar band to perform in a riff on *The Tempest;* they mounted an avant-garde production of *Frankenstein* in which the monster consults with a sex therapist; they had twenty-five local child actors perform a play about Beethoven that they staged with much of the cast wading in the waters of Lake Michigan.

If Diane's parents ever wondered during those years of living and directing on a shoestring budget when she would become financially secure, they never mentioned it. "They didn't ever say to me, 'Get real in your life,'" Diane said. "When I think about it, I didn't have that pressure." When Diane was in New York, she worked for a family friend in a temp job; she lived at home and did not have to pay rent. "And I was pursuing what I loved," she said. "My parents were always supportive. There was a reverence for theater. And my father had this spirit; my mother—she was a rebel. She got on a boat and married my father at a time when her family didn't recognize the marriage. They both understood what it was to be a free spirit, to choose the uncharted path. If you looked at them, they weren't outward hippies—they didn't present that way. But they were ahead of their time."

Janet and Stephen had left home at around fourteen; for a long time, Diane, by contrast, seemed destined never to leave. At first, Randy moved into the studio next door, so close that Teruko would bang a spoon on the wall to let him know he was welcome to join them for dinner. But eventually he moved in; the two lived in the Paulus family's two-bedroom apartment into their late twenties out of financial necessity.

As Diane had been directing child actors in the waters of Lake Michigan and doing temp work, Stephen had already, for a decade, been working his way up through the ranks at WCBS-TV, where, with the help of his father, he had first landed a job as a researcher in the news department after completing a graduate program in journalism at New York University (following his undergraduate years at the University of Minnesota).

For Laurence, that office, where he and his son overlapped for just a year, had been a source of some dissatisfaction: He lost whatever joy he felt in the job when the kind of live television arts programming he most loved fell out of favor in the seventies, and he was moved into middle management in the finance department. For many years, Laurence had sought professional fulfillment outside of the office; he was a man long on ideas but short on follow-through. He took the family on tours of many failing movie theaters within an hour or two of the city, as he contemplated the possibility of taking one over; he purchased acres of land in the Berkshires with the hope of converting it into some kind of "Hansel and Gretel" experience for paying guests. He loved New York City auctions and sometimes came home with supplies such as fifty pounds of glycerin soap or Ceylon tea. It fell to Teruko to package them beautifully for gifts they would hand out over the course of years. Caught up in the seventies-era mania for Chinese food, Laurence served his family elaborate banquets that took hours of preparation. "He was the chef, and it was the family joke that he'd make a mess in the kitchen, use thousands of dishes—and my mother cleaned it all up," said Diane. "It was a metaphor."

Stephen had considerably less exuberance than his father but a better grasp of corporate culture; from his mother, he seemed to have internalized an unrelenting work ethic. He never said no to a shift, realizing, after a few promotions in fewer than two years, that he was already earning more at WCBS-TV than his father had at the time of his retirement. On big news nights, he slept at the office; he biked downtown with bagels for the crew on Christmas. In 1991, Stephen was part of the team that founded NY1, then a novel twenty-four-hour channel devoted exclusively to covering New York, one of the first of a handful of local twenty-four-hour news channels in the country. (Among its lighter features that year: coverage of a production of Shakespeare's *Twelfth Night* being produced in an Upper West Side community garden.) He started out as news director; two years later, he was running the channel, which became one of the most-watched sources of local news in the city. The news format was successful enough that Paulus spent much of his time negotiating consulting deals with international media companies hoping to duplicate the success of a brand that quickly became a New York institution. "When Steve Paulus is around, big things tend to happen," wrote *Crain's* in 1995 when it included him in its annual roundup of forty influential New Yorkers under the age of forty.

A high-profile media figure in the city, he sometimes found himself unhappily in the spotlight: In 2010, a dismissed reporter, Adele Sammarco, filed sexual harassment and discrimination charges against several defendants, including Stephen (who was said to have commented inappropriately about female coworkers' bodies); the jury ultimately deemed Sammarco's lawsuit groundless after less than an hour of deliberation. When Stephen exited in 2016, he was recognized for his influence in New York media, having led sixteen local news channels and overseen the debut of New York's first entirely Spanish-language twenty-four-hour news channel.

When Diane was in her twenties, building a scrappy experimental theater company, Stephen, a rising executive, sometimes made it clear he found it entertaining that his sister, for all her credentials and Phi Beta Kappa gold stars, was still living at home along with her boyfriend. "I remember my brother used to joke, 'Oh, Diane is the one who went to Harvard, but she and Randy are going to have some house with the laundry in the yard, and the kids will be doing experimental theater,'" Diane said. "Any opportunity my brother had to rub experimental theater in my face."

The years Diane spent living with her parents through much of her twenties turned out to be time for which she'd be grateful the rest of her life. By the time Diane was twenty-seven, she would not have moved out even if she could have afforded it: That year, 1993, Teruko fell ill with the cancer that would take her life two years later, at age sixty-seven.

In the last months of her life, Teruko opened up to her children in ways she never had before. For the first time in their conversations, her children could detect hints of regret. "Try to spend time with your female friends," she told Janet, who heard in that wish her mother's own disappointment at not having done more of it. So much of her life was given to her tireless hours of work and her children, as well as Laurence, and so few women she knew understood her; she had Japanese friends in New York, but none had married American men. To Diane, Teruko made it clear that although she loved her husband, she also felt she had lost something of herself along the way.

"I think my mother, in her own way, was radical in her courage," Diane said. "But the liberation of my mother, as a woman, was not part of her journey. She was seeing that in my sister, and myself—and I think she appreciated what we could become, and the voices we could express ourselves." Teruko rarely expressed her feelings, but through the music of the household, perhaps, she could privately explore the depth of her emotions.

Teruko loved Puccini, Janet said, and played nothing more often than *Madame Butterfly*, one of her favorite operas—the story of a selfless Japanese woman who denies herself everything out of love for her child and the American Navy lieutenant who forsakes her, despite his promises for their future.

Diane completed her MFA at Columbia University's School of the Arts in 1997. Around that time, she and Randy started working on an immersive adaptation of *A Midsummer's Night Dream* that would celebrate the glitz and community groove of seventies-era dance floors. Diane had grown up seeing formal ballets from the inside, attending operas with her parents, and taking in countless Broadway shows; but she and Randy had also been club kids, regulars at Studio 54, the Roxy Roller Rink, and Danceteria, where they danced to Run-D.M.C. at the band's first LP release party and watched, on other nights, performance artists like Karen Finley and John Sex push the boundaries of what could be considered entertainment or art. Now they wanted to make a production of Shakespeare that would feel like a mad party, glamorous, a little debauched, and unleashed. Having already made a rock version of *The Tempest,* they approached *A Midsummer's Night Dream* with the attitude of upstarts and meddlers: Why couldn't a Shakespeare production have the energy and draw of a nightclub? Randy originated the production, called *The Donkey Show,* at a tiny venue downtown. But when Jordan Roth, an ambitious rising producer, took on the project, the show found a bigger space; more seasoned sound and lighting designers came on board, established professionals who appreciated immediately that Diane's standards were at least as high as theirs, said Randy. The show became the breakthrough project that first established Diane as an original talent with a clear future. "If music be the food of love, 'The Donkey Show' is a bowl of jalapeños," wrote Peter Marks in *The New York Times* when the show opened at the club El Flamingo in Chelsea in 1999. The play, which challenged critics with its gyrating dancers and nominally Shakespearean plot, was intended to have a six-week run; instead, it ran for six years, with additional productions opening in London and Edinburgh, then again at the American Repertory Theater, where it ran for close to a decade.

When *The Donkey Show* opened in New York, Diane was thirty-three. The success she achieved followed many years of work without widespread recognition. One could argue that she ultimately succeeded because her parents were not preoccupied with achievement; if anything, the opposite. They had given her, as well as Janet, permission to pursue the arts rather than the security of a more conventional career track. They provided a literal home for Diane's creative ambition, a haven in which she could continue to move forward, even when her future was far from certain. Right after graduating from graduate school, she considered taking a prestigious directing fellowship rather than diving into rehearsing what would ultimately become *The Donkey Show*. "Follow your heart," her mentor, the director Anne Bogart, told her. "Therein treasure lies." She could follow that advice from Bogart because she'd been hearing some version of it all of her life.

Stephen and Janet never performed as Romeo and Juliet at Shakespeare in the Park; but Diane, now the artistic director of the American Repertory Theater at Harvard, went on to direct the revival of *Hair* at the Public Theater's Delacorte Theater in Central Park; it was a show she'd loved ever since her brother brought her to see it at the Ziegfeld as a child. She directed many other large-scale musicals on Broadway, often shows with women—their suffering and their strength—at the center of the stories: *Waitress,* which featured an ensemble of working-class women searching for self-determination, made history with an all-female creative team, and *Jagged Little Pill,* the most successful show in the history of the ART, featured a seemingly powerful but struggling mother trying to help keep her family together in the face of traumas of the past and the present.

Diane is one of only six women who have won a Tony for best direction of a musical—in her case for a Broadway revival of *Pippin* in which she cast Patina Miller, a woman, in the role of the Leading Player so memorably once played by Ben Vereen. "For me, the theme of *Pippin* is this: How far do we go to be extraordinary in our lives?" Paulus told an interviewer for the Denver Center for the Performing Arts shortly before the show, which featured trapeze acts and tightrope walks, opened on Broadway in 2013. "Right now, that is such a relevant question—more than ever. Just how far do we push ourselves? What is glory? What is it to be extraordinary and what are the choices that we make in our lives?"

At the Tony Award ceremony, when she accepted her award, she dedicated her win to her own family, her husband and her daughters; but she first thanked Stephen, who was sitting in the audience, then Janet, then her parents, who, she said, her voice just barely cracking, "gave me the best gift a daughter could ever hope for: the encouragement to do what you love with your life, which for me was the theater."

Although Diane Paulus is best-known as a Broadway director, for many years she also paid tribute to her mother's greatest passion, opera, directing *The Magic Flute* for the Canadian Opera Company, and productions of *Orfeo, Don Giovanni,* and *Le Nozze di Figaro* at the Chicago Opera Theater.

Stephen watched as his baby sister's reputation grew to a level of fame that surprised him—that even surpassed his own renown in the town that he covered. He recalled attending a production of the Metropolitan Opera one summer in Prospect Park for which NY1 was a media sponsor. A friend who worked at the Metropolitan Opera spotted him sitting in the VIP section and approached him, eager to introduce him to the general manager, Peter Gelb, because, Stephen assumed, he was an important figure at the channel sponsoring the event. When the two walked over to Gelb, his friend said: "Peter, I'd like to introduce you to Steve, who is Diane Paulus's brother."

"So at that point I realized I was my sister's brother," said Stephen with a mix of resignation and pride. "That was my claim to fame."

LAURENCE PAULUS WAS A CLASSIC enterprising parent, a man who had strong philosophies of child-rearing and went to great lengths to give his children exposure to greatness.

But the enterprising parent sometimes overlaps with another recurring type whom I encountered in my reporting: the thwarted parent. Those parents are not necessarily defeated or discouraging; on the contrary, they are parents whose dreams, perhaps especially in notably challenging fields, have not been realized, and who gladly try to clear from their own children's path whatever obstacles may have been in their own way at an earlier time.

Another influential New York City family—the Wassersteins—was headed by a mother who was a well-recognized character in her own right, and yet a woman whose ambitions were, at least to her mind, unfairly stymied. Lola Wasserstein was the mother of five children, among them three who were especially high-achieving: Bruce Wasserstein, one of the most influential financiers of his time, chairman and CEO of the financial services firm Lazard; Wendy Wasserstein, the playwright; and Sandra Wasserstein Meyer, a pioneering female marketing executive.

Lola had high hopes for her own fame—she was desperate as a young woman to be a dancer, all her children knew. Her father, an intellectual and a high school principal, forbade it. Lola made sure that her own daughters took dance lessons, and when they did, Lola remained by the window, looking inside the class, not to admire her daughters' form but to practice the various moves herself. A tiny woman, she went through life carrying herself with the bearing of a star—leather pants and leotards well into her seventies, a chest-out posture—spending many of her waking hours in dance classes with

Broadway professionals, defying and denying her age comedically when the subject arose. (Wendy's first play, in fact, was produced at Playwrights Horizons because Lola ran into someone she knew from a dance studio who had recently started working at that theater.)

That three of Lola's five children went on to have not just big careers but very visible, public careers would have gratified her. Melissa Levis, a daughter of Georgette Wasserstein Levis, another Wasserstein daughter, who ran an inn on a historic estate in Vermont, recalled her grandmother as loving, even if that love was shot through with expectation. "My grandparents valued conditional love," said Melissa, who was a songwriter and acclaimed children's performer in New York before she took over her parents' inn in Vermont, the Wilburton. "It was not unconditional love." She could imagine the pressure that her mother's generation felt from Lola, because even she also felt pressure to excel at her grandmother's home, a sprawling apartment whose walls were covered with framed news clippings about Bruce and Wendy. Melissa could easily conjure up her late grandmother's voice, a nonstop patter whose message was always on point: *Do better.* "You're on JV crew? Great, what about varsity? You're in the play? You have to be the lead. You have one baby? When's the next one? Our family valued accomplishments," she said. "It was so clear." That pressure was accompanied, however, by the exuberant pride her grandmother felt in whatever it was she did do—whether it was a middle school chorus concert in Vermont, where her grandmother would travel from Manhattan to see Melissa sing, or a cabaret performance downtown. For her grandmother, living vicariously also meant showing up at every concert, every recital, every dance performance.

The playwright Tony Kushner, whose sister, Lesley, became an artist and whose brother, Eric, was the principal horn at the Vienna Symphony Orchestra for some thirty years, grew up with a mother whose own thwarted ambitions were core to their family history. His mother, Sylvia Kushner, was a star bassoonist, a talent who was, at age eighteen, one of the first women to hold a principal chair in a major orchestra, the New York City Opera orchestra. When her oldest child, Lesley, struggled as a toddler, Sylvia believed a psychotherapist who suggested to her that Sylvia's demanding performance schedule, which often took her from their home, was the problem. Lesley's tantrums, they later realized, were caused by a significant hearing loss, rather than disposition or delays; but, by then, the family had moved to Lake Charles, Louisiana, where William Kushner, Tony's father, who had played clarinet for the Houston and New Orleans Symphonies, took over his father's lumber business so Sylvia could attend to the children. Tony recalled his mother as having "a great sense of loss and a wounded professional pride," as he told *The Guardian* in 2022. In an earlier interview in *The New Yorker*, he expressed similar sentiments: "There was a sense of the world not having gotten her, and not appreciated her. She was furious about it." She was a thwarted parent, and her ambition for her children poured over and into them: She or her husband drove Eric from Lake Charles to horn lessons in New Orleans once a week—four hours each way. Tony, too, felt the depth of her desire for him to achieve. "It was this huge thing to her if I succeeded," he said in *The New Yorker* profile. "I don't recall her ever lavishing a huge amount of praise on me, but I remember the thrill in her voice when I told her I had won some debate tournament, or when I got an agent, or when I got a grant or a good review or any indication that what she expected—which was that I would be a successful artist—was going to come to pass. She would simply say, 'Go! Go! Go!' in a crescendo of pitch and volume, when I brought her good news."

His mother channeled much of her emotion into local theater in Lake Charles, where she was, at least in Tony's recollections, "an amateur actress of considerable emotional depth and power, a real tragedienne," as he wrote in *The Kenyon Review* in 1997. "I do theater because my mother did it." His father, too, never stopped making music, conducting orchestras in Alexandria, Louisiana, and in Lake Charles; but both parents, he said, "had a feeling of not having been successful and also not being fully appreciated for their gifts and talents," Kushner told me.

One sees the power of a thwarted parent in Jeannine Groff, a national champion twirler and scholarship winner once hell-bent on becoming a doctor; one sees it in Ying Wei, the mother of the Chen siblings, a woman who had longed to receive an education—and to play an instrument—but was never given the chance; and in Ellis Marsalis, the father to Wynton and his talented siblings. "At the time Wynton was growing up, I still had a lot of anxiety about going to New York," Ellis Marsalis told Wynton's biographer, Leslie Gourse. Instead, he had remained in New Orleans, where he was widely respected but never reached the level of fame achieved by peers who headed north. Gourse asked him if he thought Wynton had "set out to aim higher" in heading to New York, having observed his father's frustration. "Was Wynton trying to fulfill his father's dream?" Gourse asked Ellis in one interview. The response: "'Could be,' Ellis said, nodding slowly. 'It could be.'"

One might even see the thwarted parent in Patrick Brontë, whom the essayist Elizabeth Hardwick summed up, in an essay on the Brontë sisters published in *The New York Review of Books* in 1972, as a "failed writer." Brontë was a published local author whose inspiration for writing was largely evangelical in nature. He had a clear passion for great literature, however, and he let his daughters read widely at a time when that was unfashionable, wrote Brontë biographer Juliet Barker. In one of his works, he wrote of the "real, indescribable pleasure" that he felt when writing poetry. Brontë accomplished great things in the service of his community—but he did not receive the accolades and renown that might have once been dreamed of by a man with the ambition to rise from poverty in Ireland to success at Cambridge.

A friend of mine and her mother used to reflect on a concept that speaks to the love and respect that gets mingled with ambition in some of these high-achieving siblings: the finishing child. The finishing child is the one who fulfills the ambitions of parents who had dreams and talent but for whom the timing was not right or the world was not ready. That child and their parents might not even be conscious of the dynamic; and yet that child's fulfillment, once achieved, could feel like the answer to an earlier call, an evolution that has the feel of a fate.

Maybe even more than a parent who has achieved great public accomplishments, a thwarted parent might provide tremendous motivation. Both Diane and Stephen Paulus operated in areas in which their father had run up against limitations, either his own or those set by changing circumstances; both children eventually exceeded even the expectations he might have originally had for himself in those respective fields.

In 2002, Diane, then thirty-six, wrote and staged a production called *Swimming with Watermelons*. A play with a sweet sensibility, it was based on her parents' love affair in Japan and produced at the Vineyard Theatre in Lower Manhattan. Her mother had died years earlier; her father was still alive. Going through her parents' souvenirs and mementos, Diane had stumbled on an old newspaper clipping that featured a photo of her father with a script in his hand, his legs crossed. "And it was like, 'Laurence Paulus tears his hair out as the show must go on tonight,'" she said, recalling a caption in the article about the play he was directing in Tokyo. "I knew my father loved the theater—my father took me to the theater my whole life. But it was a light-bulb moment when I looked at the clipping. And I thought: Now I understand why I do what I do." She related strongly to the focused, absorbed, maybe harried person in the photo, whom she saw as a colleague in that moment: "I thought, *Okay, I've been there.*"

Randy finds Diane's light-bulb moment baffling, because as he remembers it, Diane's father often—even relentlessly—told stories about his days in the theater, describing at length the productions, the size of the audiences, the names of the actors. "The high point of his life was directing in World War II," Randy said. "He would talk about it over and over again."

Laurence's stories about his days in the theater were always filled with humor and told with smiles. Randy thought that when Diane saw that clip, she was comprehending something else for the first time—the extent of her father's longing for a different path. She was recognizing in her father the thwarted parent who coexisted with the joyful father. Seeing him so clearly inhabiting that role—in a way that reminded her so much of herself—she could sense all he had lost when he returned to the United States; he never managed to return to the career he once loved.

Swimming with Watermelons was reviewed as "cheerful and endearingly performed" in *Variety,* but the play also hinted at some of the painful dynamics in the couple's marriage, including its essential power differential at its outset. For Diane, the play was an opportunity to feel close to her parents and to honor their history, but the project was also emotionally challenging, as she was making art about her mother even as she still mourned her loss.

Diane Paulus, said Rachel Murdy, one of the actors in the original production of *The Donkey Show,* is a strong director, usually poker-faced in sharing her impressions and always in control of her emotions. But Murdy recalled giving a workshop performance of *Swimming with Watermelons* before it launched at the Vineyard and seeing Diane weeping, leaning against a wall for support as she sobbed in grief.

"I was overcome with the pain of wishing I could speak to my mother and ask her all the questions I never asked," Diane said. "You never realize how much you don't know about your parents until it's too late."

Teruko was neither an enterprising parent nor a thwarted one; perhaps she thought of herself, toward the end, as a cautionary tale, a parent who maybe lost herself along the way as she raised children who themselves felt empowered. But she was clearly also an over-comer, a literal survivor, and a quietly extraordinary woman whose strength was part of what inspired—maybe even enabled—Diane and

Janet to become artists. "Diane worshipped her mother," said Randy. Added Janet, "We all did."

Laurence's explicit forms of enterprise parenting—the maps, the composers and lines of Shakespeare he had his children memorize— were bits of showmanship that reinforced the children's love of the arts. But why they excelled in their respective fields, all three children believe, was because of Teruko. What she sewed in her makeshift workshop was fine art of a kind; what they saw, every day, was the patience it took to keep sewing one more tablecloth, to listen, around a dining-room table, to what her children had to say, and to watch, without desire, as her children gravitated to their callings.

Diane's father would end his days in a retirement home, singing old show tunes and flirting with fellow residents. But a few years before he died, he was able to attend one of the performances of *Swimming with Watermelons* at the Vineyard Theatre. When Diane announced his presence, the audience gave him a standing ovation of the kind that Diane knew he had craved all his life.

TEN

Openness

In 1861, Edward Emerson—the son of Ralph Waldo Emerson, the American essayist, philosopher, and abolitionist—was attending a progressive private school in Concord, Massachusetts, with Wilkie and Robertson James, the two younger brothers of two of the most famous thinkers in the twentieth century: the novelist Henry James (author of, among other great works, *The American* and *The Portrait of a Lady*) and William James, the oldest sibling, a philosopher and thinker whose work formed the basis for the modern field of psychology.

Over one spring vacation, Edward, then seventeen years old, visited the James family at their home in Newport and captured, in a letter, the scene around the Jameses' dinner table, which was as extraordinary as the brothers themselves:

> "The adipose and affectionate Wilkie," as his father called him, would say something and be instantly corrected by the little cock-sparrow Bob, the youngest, but good-naturedly defend his statement, and then Henry (junior) would emerge from his

silence in defense of Wilkie. Then Bob would be more imperti-
nently insistent, and Mr. James would advance as Moderator,
and William, the eldest, would join in. The voice of the Moderator presently would be drowned by the combatants, and he
soon came down vigorously into the arena, and when in the
excited argument the dinner knives might not be absent from
eagerly gesticulating hands, dear Mrs. James, more conventional, but bright as well as motherly, would look at me, laugh-
ingly reassuring, saying, "Don't be disturbed, Edward; they
won't stab each other. This is usual when the boys come home."
And the quiet little sister ate her dinner, smiling, close to the
combatants.

The description of that family dinner sounds remarkably like the
freewheeling Foer family dinners, meals conducted in a spirit of open
inquiry in which no topic was off-limits. It also calls to mind the spirit
of dinner-table conversation shared by the Emanuel brothers: Rahm
Emanuel, who served as mayor of Chicago and as the US ambassador
to Japan; Ari Emanuel, chief executive of the influential sports and
entertainment company Endeavor; and Ezekiel Emanuel, a prominent bioethicist and professor at the University of Pennsylvania. "You
had to get ready for dinner conversations at our house," Rahm told *The
Washingtonian* in 2008. "You either came ready or you got shut out."
He's also described those meals as "gladiatorial" in energy.

The Emanuels' father, Benjamin Emanuel, an Israeli-born social
justice–minded pediatrician, welcomed challenging ideas, bequeath-
ing two opposing thoughts to his sons: "One is to always challenge
authority; the other is to always respect it," Rahm recalled.

Henry James, Sr., also prized irreverence and endeavored to impress it upon his children. In her diaries, his daughter, Alice, recalled
her father demanding that whoever spoke at his funeral should "say
only this: 'Here lies a man who has thought, all his life, that the cere-

monies attending birth, marriage and death were all damned non-sense.' Don't let him say a word more." Henry Sr.'s "radical, sometimes naïve questioning of established norms, and a willingness to go lightly into zones traditionally characterized by seriousness" were qualities that his children adopted for themselves, especially William, writes Jane F. Thrailkill in her book, *Philosophical Siblings: Varieties of Playful Experience in Alice, William, and Henry James.*

It takes considerable irreverence, but also some ego, to dismiss as nonsense the ceremonies widely considered to be not just norms but nearly sacred. Brilliant enough to attract the friendship of men like Emerson and the writer Thomas Carlyle, Henry James, Sr., who was born into great wealth, was nothing if not ambitious, writing numerous books that set out to capture his deepest philosophical thoughts on the ideal relationship between humans and God. Like Patrick Brontë, he never received serious attention for his books, which did not stop him from continuing to publish them. He found perhaps more fulfillment in what he perceived as his other calling, which was educating his talented children. Having been raised in a stifling and authoritarian environment, he wanted his own children to feel empowered to express themselves.

"Curiosity and irreverence go together," the progressive activist Saul Alinsky wrote in his influential 1971 book, *Rules for Radicals: A Pragmatic Primer for Realistic Radicals.* "Curiosity asks: Is this true?" To the curious, he continued, nothing is sacred; everything could be questioned.

A questioning spirit, a powerful curiosity—those are common qualities in many trailblazers and innovators and creatives of note. Those are also the qualities, Julia Leonard told me, that are increasingly recognized in the psychology community as vital to cultivate in children, especially following the pandemic, as part of an emerging emphasis on what she called "learning to learn" skills. "Those are some of the skills that make you open to learning, that make learning almost frictionless, because it's just part of who you are, in the sense that you are open to new experiences, curious about them," she said. "Like if you're watching children, those are the signs of vitality—the interest in being part of the world."

Curiosity is also a key facet of the personality trait that is closely associated with creativity: Openness to experience.

"As neurotics can be used as exemplars of high scorers on the dimension of Neuroticism, so artists can be considered prime examples of individuals high in Openness to Experience," wrote Robert McCrae and Paul Costa in 1997 in their influential chapter in the *Handbook of Personality Psychology* on what are considered the "big five" personality traits: neuroticism, extroversion, conscientiousness, agreeability, and openness to experience. Those who are open to experience—those who possess what's often referred to simply as "openness"—might be more easily moved by art or music, which does not mean they are necessarily sophisticated or highly educated (though they may be predisposed to pursue education). They are more likely to be eager to travel, to delve into rich, novel experiences. They have, as McRae and Costa wrote, "an insatiable curiosity, as if they retained into adulthood the child's wonder at the world. And they are unorthodox, free-thinking, and prone to flout convention."

Openness to experience captures in so many ways the intellectual curiosity shared by people as different as Henry James, Sr., and Benjamin Emanuel. It is likely a quality shared by many of those who make the leap to emigrate, such as Benjamin Emanuel and the Chens and Amalia Murguía—it is so much safer to stay home than to take the risk of starting over in a foreign country, to be open to the new experience of an entirely new culture. Openness captures so much of the personality of Teruko Paulus, an immigrant and a creative, and Laurence Paulus, who shared her love of culture and the arts.

Openness to experience can be described as "the drive for cognitive exploration of inner experience," in the words of cognitive scientist

Scott Barry Kaufman. If conscientiousness is a personality trait associated with planning, with thinking ahead, openness is a personality trait associated with being intensely present in the moment, should that moment be transporting, or with daydreaming amid the mundane. Those who are open are not so much people who let their interiority out—in the sense of being transparent or overly honest—but rather who let the outside world rush in. One of the markers of people who are high in openness, across cultures, is being prone to chills from music or theater or unexpected beauty.

Openness is associated with high measures of happiness and even exceptional adaptability to stress—both of which Ying hoped she could offer her children through music. That personality trait, some research indicates, is also strongly linked to some early predictors of achievement. Intellectual curiosity and creativity were found to be more important in predicting differences in students' reading and mathematical abilities than qualities such as conscientiousness and discipline, according to research published in 2019.

Could parents who are high in openness to experience—who want their children to share their passion for travel and the arts, to feel those chills, to benefit from that hungry curiosity—somehow instill that trait in their children? One extended project in Turkey gave some teachers pedagogical tools to pique students' curiosity about assigned topics, helping them devise methods that relied on elements of mystery, humor, and playfulness. The project found that the students who benefited from those teachers' approaches, compared to the controls, did exhibit more curiosity about learning, as measured by their willingness to trade tokens, at the end of the experiment, for booklets they'd been promised would teach them new facts about various topics.

Sule Alan, who conducted the research and is now a professor of economics and behavioral science at Cornell University, said she does not see her work as attempting to change childrens' personalities;

instead, she sees her efforts as designed to reinvigorate in children a quality that is innate. While she is working with children, she is often trying to rebuild that curiosity, which she believes is too often quashed by rote pedagogy. "We want to form the habit of questioning, the habit of not believing everything that is given to you," she told me. "And that is going to make you exploratory and willing to take risks—there's a new thing to grab, go grab it!"

Those teachers in her study received extensive training and materials, the likes of which no parent is likely to obtain. But parents who want to cultivate this particular personality trait in their children and have the luxury of the resources might encourage a particular choice when their children are college-age: travel. Researchers in Germany, for a study published in 2013, tracked Europe-based university students who traveled to other countries in Europe to study and found that long-term travel enhanced not just their emotional stability—their ability to weather a challenge—but also their openness to experience, especially when the trip transcended tourism, entailing real, prolonged engagement with another culture. The more the social networks of that person changed and expanded over time, the more their openness to experience grew, as the power of connecting with people from different environments, and with different values, supercharged whatever emotional growth they were already undergoing during the crucial developmental phase of those late teens and early twenties.

Personality researchers caution that the impacts of cultural exposure on traits are not fully understood (and—as is true of the effects of so many life experiences or deliberate interventions—do not necessarily last). But a deep love of travel, one indication of openness, runs through many of the families of this book. The James children's childhoods were characterized by schooling in cities across Europe. Travel, for the Emanuels, was a powerful family value. "For my father, travel was not a luxury," wrote Ezekiel Emanuel in his memoir, *Brothers Emanuel.* "It was, he believed, absolutely necessary for an understanding of the world, and of oneself." Both Lauren and Adam Groff traveled abroad for a year before attending college, experiences that they both describe as formative. Laurence Paulus not only traveled extensively but raised children who do the same: Stephen was educated in Switzerland and sailed the oceans, Diane's work has been produced on four continents, and Janet has lived abroad for much of her adult life—she spent five years in Spain before moving permanently to Mexico.

WHEN PARENTS DO ASPIRE TO foster in their children qualities such as curiosity or wonder (or grittiness or conscientiousness), they are essentially trying to shape their children's character in certain ways they deem desirable or that they believe will help their children succeed. Should they blame themselves for failing to instill those qualities, their sense of guilt rests on the shaky premise that they had significant power to shape their children's personalities in the first place.

What does change personality, if anything? Walter Mischel threw a bomb into the field of personality research in 1968, when he published a book called *Personality and Assessment.* This book challenged the assumption, previously long held by a majority of researchers in the field, that personality could be summed up by traits that essentially remained stable across time and in varying situations. In 1973, he published another influential paper, suggesting that the field of psychology, in trying to understand human behavior and performance in terms of personality traits, was overlooking the context and emotional states of the humans involved. People behaved inconsistently, changing their behavior depending on their circumstances, argued Mischel. He based his proposition on the reanalysis of dozens of studies; but his own life experience also bore that theory out. As a child, Mischel had watched his own mother, a neurasthenic lady of leisure in Vienna, become a hardworking waitress in New York, where the family had fled, all but penniless, after the Nazi occupation. His father, by contrast, a powerful man with an authoritative air in Vienna, languished in New York, never recovering from the decline in the family's life circumstances.

Mischel was also famous for being the scientist who pioneered the marshmallow test—research that followed, over years, the benefits that accrued to children who were able to delay gratification at a young age. The young test subjects, most of them around four years old, were told that they would be rewarded with a second marshmallow if they could put off eating the first one. Those who delayed gratification fared better in terms of education, health, even marital satisfaction, decades later. (Other studies, however, have failed to replicate the long-term findings, and critics have pointed to problems with the homogeneity of Mischel's cohort, which included many offspring of Stanford academics.) Mischel, like Duckworth, consulted with the KIPP schools, and the pursuit of non-cognitive skills such as discipline, patience, and grit captured the imagination of American educators.

Grit speaks to a deeply American belief—the wishful thinking that with hard work and enough passion, anyone can overcome disadvantage. Openness to experience and curiosity, by contrast, has perhaps long had a whiff of something more suspect—intellectualism, rebellion, even a preoccupation with questioning, subversion, or seeking difference. The pursuits of those open to experience might even seem self-indulgent. Henry James, Sr.'s closest friend, the great American essayist Ralph Waldo Emerson, could have been directly critiquing his dear friend's choices to travel and take his children abroad, again and again, for their education, in an essay he wrote about wanderlust. "For the most part, only the light characters travel," wrote Emerson, who himself traveled widely. "Who are you that have no task to keep you at home?"

Now newly valued, openness has been the focus of researchers in pursuit of what's known as a "nudge," a low-stakes, inexpensive intervention that might yield surprisingly dramatic results. In another study on openness, published in 2018, researchers tried to encourage those low in openness to expand their horizons. Students at a university were asked to write essays in response to assignments promoting introspection, or to explore aesthetics, ideas, and feelings. The results disappointed them: Their work had no effect on their openness, albeit after only five days of the experiment (the controls of which simply wrote about the details of their day).

It was not the short-term duration of the study that doomed it to fail: Those who were already high in openness did see increases in those levels, as they were likely inclined to find the task energizing. But those who were not high in openness at the outset saw no change whatsoever. As is true of so many traits we deem desirable, simply encouraging or even finding ways for a young person to adopt them is not likely to effect a change; if anything, researchers often find, such efforts can be demoralizing. Introverts, for example, asked to engage in behavior that requires extroversion, might initially feel energized, but then experience "depleted levels of vitality later on," wrote the authors of a study that ran in 2020 in the *Journal of Research in Personality*. Ultimately, being asked to perform in a way that did not feel natural only backfired.

In 1890, Henry James, Sr.'s oldest son, William James, published the seminal work *The Principles of Psychology*, in which he expressed some of the profound and original ideas for which he is best known—popularizing the notion of the "stream of consciousness," as well as the theory that different parts of the brain correspond to different aspects of thought, emotion, and feeling. He also, famously, weighed in on that enduring question of personality, a matter of ongoing debate and discussion in the field even now: Is personality stable? Can it change over time? Does it change in

predictable ways? Can it be changed by design? Under what circumstances should it be?

James argued in *The Principles of Psychology* that who people become is a function of the habits that shape them over the course of
their lives, habits that develop in response to the environment in
which they are raised and the daily choices they make every day within
that environment. He wrote that that provided an essential and even
freeing stability, one that was, he argued, restricting but essential, lest
we all live in a world of unpredictable upheavals. "It is well for the
world that in most of us," he wrote, "by the age of thirty, the character
has set like plaster, and will never soften again."

Henry Sr., for all his child-like enthusiasms, his devotion to his
children, and his openness to experience, had fairly rigid ideas about
who he wanted his children to be—especially William, the oldest, for
whom he had, perhaps, the highest expectations. When William, who
clearly shared his father's openness, was drawn to painting as a vocation, his father deliberately interrupted his training by picking up and
moving the family to Europe for one of their many transatlantic mi-
grations. At his father's urging, William eventually enrolled in medical school, a pursuit for which he was temperamentally so ill-suited
that it marked the beginning of a period of intense psychological
misery—debilitating depression—that he called his era of "soul-
suffering."

Only in the last years of his father's life did William James, by
then in his thirties, began seriously indulging the more abstract and
philosophical questions that preoccupied his thoughts, focusing more
on those early concepts of psychology. A decade after his father's
death, he published a primer on psychology based on the *Principles*,
in which he coined an aphorism suggesting that individuals are in
charge of deciding the kind of people they will become: "Sow an action, and you reap a habit. Sow a habit and you reap a character; sow
a character and reap a destiny."

There is often a grim undertone to the way James talks about habit, as if his core life's struggle had been to establish his own habits in place of those that had been imposed on him. Habit, he wrote, "dooms us all to fight out the battle of life upon the lines of our nurture or our early choice, and to make the best of a pursuit that disagrees, because there is no other for which we are fitted, and it is too late to begin again." As a tool for developing the self, habits may work, but they don't seem to travel well; one person's habit is another's hell. James's famous aphorism might be best understood as a philosophy not for the parent but the child: Mold your own character. Choose your own habits. Sow yourself.

ELEVEN

The Wojcickis

I magine a mother, a striver, someone who got herself an education despite her parents' low expectations for her, someone who'd be damned if her expectations for her own three daughters would be anything but high. When her daughters wrote papers for their competitive high school in Palo Alto, their mother, who taught journalism at a nearby high school, returned them in a sea of red ink. You can turn this in and get a C or a D, their mother would tell them, as neutrally as possible, or you could rewrite it. Maybe it really was a C or a D paper; maybe it was not. In her opinion—an opinion that mattered a lot in that house—it was not yet ready for submission. Each of her daughters would take the paper and rewrite it. The paper improved—this is a B, she would say. They would rewrite it again; they wanted the A.

This sounds like the story of a wise mother. She had high expectations, but she did not force her children to rewrite the paper; she gave them agency. Isn't the proof of her wisdom the rewriting of the paper—the eventual A?

Two of these daughters grew up to be immensely influential

women, historically significant, even; a third would be a Fulbright scholar, a tenured professor whose research has focused on the spread of HIV and childhood obesity, who speaks French, Spanish, and the African language Kiswahili.

In 2019, the youngest of those daughters, Anne Wojcicki, the founder of 23andMe, the direct-to-consumer gene-testing business, was featured on the cover of *Inc.* magazine, wearing a crisp white shirt, smiling with wide-mouthed amusement, looking heavenward, as if reflecting on a fond memory. "The Best Business Advice I Ever Got," was the coverline of the magazine. Anne, who contributed an essay on that topic, started out talking about those papers that her mother, Esther Wojcicki, edited. "I really admire how my mom teaches kids," Anne wrote in *Inc.* "She doesn't use punishment. It's 'You turned this in, and it was bad. Learn.'"

Anne Wojcicki said she employs that same method of feedback in her business, which was intended to revolutionize genetic testing and has been used by some fifteen million customers.

Somewhere out there, however, there might be some child with a temperament different from that of Anne Wojcicki, someone else whose mother told her she would get a D if she turned in her paper as it was. That child might collapse a little internally, might bridle at her mother's opinion; she might rip up the paper or decide she was just a bad writer and come to loathe sitting down to the task. Years later, in therapy, she might tell her therapist that her mother was overly involved in her schoolwork—none of her friends had mothers who critiqued their papers so heavily. That story does not surface in *Inc.* magazine; if the person is successful, she tells a different story. If she is not particularly successful in a conventional way, *Inc.* never interviews her in the first place.

Parenting advice should come with a small caveat: Don't try this at home. Or at least: Results may fail to replicate. And yet so many people have asked Esther Wojcicki for her advice over the years about how she raised her three daughters that she wrote a book about it, titled *How to Raise Successful People,* published in 2019. Esther Wojcicki, who taught journalism at Palo Alto High School for forty years, has instructed thousands of students with strong results, but as a parenting expert, she was naturally working with a very small sample size. The Wojcicki sisters could be considered outliers in so many ways—twenty-first-century marvels forged under ideal circumstances, likely with inherently exceptional high-performance material. Their success reflects the intersections of fate and location and timing and hard work and ingenuity—and yes, they likely benefited from a force of nature known to her students as "the Woj"—Esther Wojcicki.

The story of the Wojcickis starts, I would argue, like the story of the Holifields, with the loss of a young child, a loss from which would spring new resolve, all of it poured into children to come a generation later. In this case, the lost child was Esther's younger brother, a sixteen-month-old who put fistfuls of aspirin in his mouth and then swallowed. Esther's parents were Jewish immigrants from Ukraine and Siberia who settled in the Sunland neighborhood of Los Angeles. They were poor intellectuals—poor enough that sometimes Esther, then around nine but growing fast (she was taller than her parents would have liked), sometimes went hungry. When their sixteen-month-old swallowed that aspirin, her parents listened to the doctor on the phone who told them not to worry. When it became clear they did indeed have to worry, that their son was fading, they brought him to the nearest hospital. Esther was with them as they rushed to a hospital where his stomach was pumped, but which refused him further care even though he was still gravely ill. Esther and her family tried another hospital, which also turned them away (Esther believes it was because they could offer no "proof of payment"). Finally, they found a hospital where the doctors tried to save him, but by then, they couldn't.

Esther believes that her own roaring engine of agency was built that day. Never again would she assume that doctors knew what they were talking about, that the authorities held hidden knowledge. She would never again believe that anyone could protect her. She would protect herself by making her own future—she would learn everything she needed to learn to make herself invulnerable.

"I just felt like I was going to get out there and change the world," Esther Wojcicki told me in an interview at her Palo Alto home, the same one in which she raised her three daughters. It is the kind of statement that, in another context, might be conveyed with a hint of rue—after all, who really does change the world, at least in ways that the world notices? And how much control can any one human have? Esther is no more immune to suffering than anyone else: She lost a nineteen-year-old grandson in February 2024; six months later, she lost Susan, who succumbed at age fifty-six to non–small cell lung cancer.

But if Esther Wojcicki did not directly change the world, she raised daughters who did.

It would have been novel enough to raise daughters in the 1970s and 1980s who became entrepreneurs in STEM fields; Anne and Susan have, beyond that, led the way in building some of the more innovative businesses of our time—businesses that have helped shape, for better or for worse, the way we understand ourselves now.

It was no accident of fate, in Esther's telling, that her daughters were innovators. Shortly after Esther gave birth to Anne, in 1973, a friend called. "I was so sorry to hear you had another girl," the friend told her. Like Esther, the friend was raised in an Orthodox Jewish home, amid a family culture that prized boys; Esther had been told, as a girl, that any money the family had would go toward sending another brother, not her, to school. Steeped in that upbringing, Esther admits, even she briefly felt a little sorry for herself upon the birth of her third daughter. But self-pity is anathema to Esther; agency is all. "'*Oh my God*,'" she recalled saying to herself in response to her friend's comment. "'*I'll show you. These girls are going to be girls like you've never seen*.' And that's what I did." In the years since, her friend has often apologized for her hurtful remark. But now Esther sometimes wonders if her daughters' ambitions might have been less bold had her friend never offended her with her honest pity. Was that insult the best thing that ever could have happened to her family?

So that was the project: Raise girls like the world has never seen. How does one go about accomplishing so grand a goal? Maybe it simply starts with being the kind of person who normalizes that sort of confidence for her children—the confidence that such a goal might be achievable. Why shouldn't Esther be the one to figure it out? Who was going to tell her she couldn't?

ESTHER DID NOT RAISE HER CHILDREN ALONE; she met her husband, Stanley Wojcicki, at the University of California, Berkeley, where she had landed a scholarship. The eventual parents of the Wojcicki sisters met like two energy particles colliding. Esther, then nineteen, was particle one, a young, high-spirited undergraduate. She was sliding down a stairway in her dorm in a cardboard box and landed with something of a crash—she was almost six feet tall—at the feet of particle two, Stan Wojcicki, a Polish refugee whose brilliance was soon apparent to her. "It wasn't easy to find someone smarter than I was," said Esther.

By the time Stan Wojcicki's daughters were old enough to understand what their father did, he was a professor of physics at Stanford University (and later would become the department chair). His greatest professional dream was to help design and secure funding for one of the most ambitious science projects of his time, the multibillion-dollar Superconducting Super Collider, known as the SSC, which would slam together particles with twenty times as much energy as any other devices like it. "What particles and forces will result when such collisions occur?" wrote Stanley Wojcicki in an editorial that ran in *The Bismarck Tribune* in 1985 and that called for federal funding of the SSC (which was ultimately abandoned on account of its exorbitant cost). "Physicists have different predictions, but the real answer is that nobody knows for sure. The only certainty is that the results will provide us with profound knowledge about our world. At a time when it is easy to despair about mankind, we all should feel renewed by this challenge." The girls' mother wanted to change the world; their father was in pursuit of unraveling its mysteries. The combined ambition in the family, from the outset, was vast, which is not to say it was self-interested. Reaching the limits of human understanding, pushing boundaries—those were the family values, as much a part of home life as washing the dishes or fighting over borrowed clothing.

Ambitious projects start small and, in Esther's case, with small children. The Wojcicki family, like the Groffs, had a place to swim right in their backyard; like the Holifield parents, Esther believed swimming was essential not just for safety but for a sense of self-efficacy. They would learn to swim because, as Esther puts it, "I made sure those kids were empowered—early on." Her daughter Susan was swimming by the time she was three; Janet was swimming, Esther said, when she was still a baby, with the help of a book Esther bought called *Teach Your Baby to Swim*. By the time each girl turned five, she was on a swim team. Anne was an early standout: At five, she won a regional championship race in butterfly, a stroke her mother had taught her even before she received coaching on the team. All three daughters could read at least a little before they got to school. "The reason I taught my kids all that stuff?" said Esther. "You have to be in control of your environment." She wanted them to be able to read not just books but road signs, the newspaper; they had to learn to count so they could be independent, bike to the deli five minutes down the hill, buy themselves an ice cream, make sure they got change. "It goes back to the death of my brother," she said. "We had no control of the environment, and it did us in."

In some ways, aspects of Esther's parenting style were not all that different from those of any other parent in her circles in the 1970s, a time when the prevailing mode was benign neglect: She wanted her children to learn to swim so that she wouldn't have to watch them in the pool, she told *Time* magazine, a reasonable decision at a time when parents' tolerance for risk was, in upper-middle-class families like the Wojcickis, far higher than it is now. Palo Alto in the seventies, when her children were young, may have been an academic community, but it had a strong holdover hippie vibe, with teachers whom students might run into at Grateful Dead concerts and parents who did not think twice about letting their young children bike around town until dinner on the weekends.

At the same time, Esther set a bar for productivity that was unusual—a Groff-like level of energy and exertion. "We were never a family that sat around, watched TV, and were like, *Oh, let's just relax,*" said Anne. "If we were watching TV, we had to get stuff done at the same time. My sisters and I went through a phase where we really loved the soap opera *Santa Barbara.* And we used to watch it every day, especially in the summer. But my mom always had me iron while I was watching it. And I made seven dollars an hour for ironing."

Anne said her mother's energy at times was trying for her daughters. "It drove me crazy," Anne said. "She'd wake me up at like seven on weekend mornings, or like six. And she'd have the vacuum cleaner and just bang into the door. And I'd be like, 'I'm sleeping! I'm sleeping!' And she'd just ignore me. She would just back her way out and then just close the door. She just totally ignored us." It was a common sight to see Esther taking apart the entrails of their car so she could repair it herself. When Esther wanted to rewire some electric circuitry in the house, she bought a book that told her how, and then did it herself, scrambling on ladders with a flashlight in one hand, the manual in the other.

Are we surprised, given Esther's modeling, that her children were exemplars of industry and enterprise—that, as young girls, they were selling lemons to the neighbors on the street, even to the neighbor from whose tree the lemons originated? It is one of the core tenets of developmental psychology that children model their behavior based on what they observe in the behavior of the adults around them, a notion that undergirds much of Julia Leonard's work. Is a certain behavior rewarded? Encouraged by other adults? Punished?

Seven years before Susan Wojcicki was born, an eventual colleague of Stan Wojcicki's at Stanford, Albert Bandura, conducted on that campus a study of children's behavior and found results that were stunning at the time. On YouTube, the site that Susan Wojcicki helped build into a massive environmental force, one can call up a smudgy, Kodachrome-tinted video of the actual experiment. In the footage, a woman in a crisp white shirt and a calf-length skirt, her hair neatly styled, plays with a blow-up doll known as the Bobo doll. At first, she takes a mallet and starts hitting the doll, without much force; then she kicks it like a ball. Finally, she crouches down on it, like a cop about to make an arrest, and pummels it with her fists (although the pummeling looks something like play—there is no viciousness on her face). As she has been acting out this fit of rage, a child has been watching from behind a glass window.

Bandura could not be sure what to expect when he subsequently put the child in front of the doll. Maybe the child would feel that his own rage had been purged by this display; or maybe he would hug the doll out of empathy. Instead, that child—whether a girl in a lacy-fronted dress or a boy with his button-down shirt neatly tucked into his pants—was far more likely to respond with force. The findings were published right around the time that violence on television first started concerning policymakers. Like other highly theatrical social psychology experiments of the time, such as the Stanford Prison Experiment, Bandura's work dramatically influenced the way people thought about the malleability of the human character, in response to forces such as modeling, and implicitly, parenting (as well as other environmental exposures).

It is an odd quirk of history that in the 1960s, a period characterized by young people's questioning of authority and adults, psychological professionals influenced by Freud were reasserting the power and influence of the ultimate adults in their lives: their parents. By contrast, the much greater amount of control that educated parents now try to exert on their children comes right at a time when new kinds of genetic research—some of it assisted originally by the collection of data by 23andMe—is filtering into journals and suggesting to the general population that one's hereditary load likely influences qualities such as personality traits.

The influential Bobo doll study is not without limits: Its research subjects were no older than five years old, their reasoning limited. How a five-year-old responds tells us little about how a ten-year-old might; and how a child responds to an unfamiliar woman playacting violence on an inflatable doll might reveal little about how that child would react following a true act of violence perpetrated by an adult on a fellow human. Science is best at observing "what happens when"; it is far less precise in determining "what happens why" and under what range of circumstances the "what" and the "when" will still play out the same way. To many behavioral geneticists, much of the research that tries to evaluate the power of parental modeling is murky. Many would argue that much of the past research on modeling, a key way that developmental psychologists imagine parents influencing their children, can only tell us so much about the extent of its influence. Researchers may see a correlation between parents who model good-natured behavior and their agreeable children, but behavioral geneticists would point out that easygoing parents are more likely to give birth to naturally easygoing kids. The parents' modeling may indeed have some impact—but how much? To complicate the matter further, children who are naturally agreeable might elicit more good-natured behavior from those around them, creating their own environment in subtle ways. The pathways to personality intersect, overlap, and double back down on one another.

In a 2023 podcast interview, Jay Shetty, the media personality and life coach, asked Susan Wojcicki to elaborate on why, in another interview, she'd called her mother an inspiration, and to talk about the ways in which her mother had been influential. Susan, looking a little amused, encouraged Shetty to give Esther a call—"I'm sure she would love to be on your show!"—and then responded, haltingly, with humor and honesty. "My mom always has something to say," Susan said. "The fact that my mom always had something to say, as a kid, was horrifying—she would embarrass me, she would complain, she would tell people how things weren't done the ways she wanted, or she would protest. She always had an opinion." If her mother complained, this was in the service of the next person's experience, Susan explained to Shetty, although that didn't seem to be her main takeaway. "She really believes in stating her opinion and not holding back and thinking for yourself," Susan said. Yes, it horrified and embarrassed her at the time, Susan reiterated—she clearly was not particularly inclined to model her own behavior on her mother's in those contexts. But "she also just taught me to think for yourself," said Susan. "If people are saying things that don't make any sense, she taught me to think for yourself and believe in yourself and believe in what you think and not just go along with what's a common thought."

In *How to Raise Successful People*, Esther lays out her guidelines for raising successful people, using the acronym TRICK, which stands for the guiding principles of Trust, Respect, Independence, Collaboration, and Kindness. In the book, Esther endearingly admits to garden-variety parenting missteps (such as reading her thirteen-year-old daughter's diary for no reason other than that she stumbled on it) and others more exotic (telling two of her daughters that she and she alone was the gifted recipient of a Volvo—a feat Esther could manage for longer than expected, because one daughter was already away at college). She also, however, takes credit for getting enough right in ways that were, it's implied, formative for her high-achieving daughters. She takes credit for modeling grit, for knowing just how much to encourage children to stick to a commitment, while also being flexible enough to let them change paths, for allowing them autonomy and respect while also being loving and caring enough to track Anne down while she was traveling in an obscure part of Russia.

Esther's book is part of a tradition of books about these kinds of high-achieving families that starts, in some distant way, with Gaskell's biography of Charlotte Brontë. Even before I started writing my own, I would have been drawn to these books, as they lay bare the household anthropologies of outliers, families that are intentional enough about success that they planned for it, a decision unusual on its face.

WAYNE BRYAN, the father of the identical twins Mike and Bob Bryan, the doubles tennis champions, both of whom are also accomplished musicians, was asked so often what his "formula" was for raising champions that he could only see providing the answer in book form as a public service. *Raising Your Child to Be a Champion in Athletics, Arts and Academics* is a slender, low-gloss, briskly paced how-to guide. The pleasure Bryan took in raising his sons is evident, even if at times his advice sounds like the kind that any well-intentioned parent might offer a friend. In advising parents on how to respond when a child loses at some competition, Bryan suggests that a parent "provide perspective" ("I once lost an important game when I was your age but now, etc. etc.") and go out for a fun dinner, like pizza, to cheer the child up. How should you entice a child to practice a sport diligently? Develop in your children what he calls Crazy Passion—and the only way to do that is to "make it Fun"—by letting the children choose the music, for example, on the way to watch a sporting event. Also? "They also have to have Fun getting pizza afterwards," he noted.

And yet, glimpses of what truly was unusual about the family mindset do occasionally pop up. In a chapter in which Bryan talks about the importance of goal setting, he advises parents to paste those goals—for writing competitions, for upcoming recitals or tennis rankings—on both the refrigerator and the bathroom mirror. My

husband and I bought our own sons a lot of postgame pizza; we certainly tried to make whatever goal they were working on fun. But I
suspect that had I posted goals on my kids' bathroom mirror, they
would have avoided brushing their teeth altogether. Even as I write
that, I feel a slight shade of guilt, of doubt—maybe I'm wrong, maybe
such postings would have reinforced some nascent enthusiasm. More
likely, they would have reinforced some sense of family values: that
achievement was so paramount that it was structurally embedded in the
daily rituals of their lives. Something in me rebels at the thought of
the boys catching a glimpse in the mirror of that list just before they
went to bed—the narrowing of the focus at a moment when they're
about to drift off to dream of things perhaps even more enticing than
their next goal.

In chapter 14 of Bryan's book, the reader hits the piece of advice
that most likely explains so much else about the Bryan household. "I
can't emphasize this point enough," he wrote. "No TV in your house!
Okay. Now that you're getting rid of your TV, get rid of all video
games too." Television and video games, he said, "will rob your child
of a chance to reach his or her true potential." Those pleasures simply
fill too many hours, he explained, hours that could otherwise be spent
on activities the children would enjoy all the more if they weren't
comparing them to the easier rewards of moving digital images.

Such advice is even less easy to follow in the era of iPhones, but
even at the time he wrote the book, in 2004, Bryan's screen-free policy
reveals less about the evils of technology than it reflects just how unusual a person Bryan was—so intent on raising champions that he
was willing to be the extreme parent, the person who does what so
many other parents likely wish they could do but can never muster the
willpower to pull off.

Bryan clearly had a drive of his own, which he applied to instilling
in his sons. His advice, throughout the book, is far from outlandish—
banish screens, set routines, which, in his family, included music practice at 7 A.M. What the book lacks is the advice that many parents

might need most—which is not how to change their children but how to change themselves into the kind of person who has the rare focus and cheerful discipline of a Wayne Bryan. If you think it's hard to change the behavior of your child, try changing yourself, Dan Belsky, an epidemiologist at Columbia University who does genomic research, once pointed out to me. "Despite our intentions, it is hard to guess the best way to change our children through our own behavior," he said. "Particularly because our own behavior is important in key ways and few people can engineer profound shifts in their own behavior."

Bryan's considerable ambition and focus for his boys is mild in comparison to that of László Polgár, another parent who went on to write a book that promised parents that they, too, could achieve extraordinary results. A Hungarian educational psychologist, Polgár raised, with great intention, three of the most successful women in the history of chess—girls who were, even by the time they were ten, established chess champions. Two of his three daughters went on to earn the title of grand master, with Judit, the youngest of the three, achieving that goal by age fifteen (breaking Bobby Fischer's record by a month). In 2002, Judit defeated the chess champion Garry Kasparov, in a win that helped change deeply entrenched perceptions of the limits of women's chess abilities—considered a proxy for a certain kind of intelligence—relative to men's. Susan, the oldest, and Judit, the youngest, at different points ranked as the number one female player in the world.

In 1989, Polgár published *Raise a Genius!*, in which he argues that "Every healthy child can be raised to be an outstanding person, in my words, a genius." Polgár liked to claim that he courted his wife, before they even married, with the promise that they would raise extraordinary children; in her own book, *Breaking Through*, Susan Polgár quotes her mother, Klara Polgár, saying that the first time she ever met László, he spoke of wanting six children and said that "he would teach them a lot with specific goals in mind and nurture them with great love." Klara went home and told her parents that although she found this son of a family friend very interesting, she could not imagine marrying him. Over time, she changed her mind.

László's book, which presumably holds the key to his own daughters' genius, is full of theories but also contradictions. He suggests that the best way to let a child find genius is to follow that child's passion. In the case of his firstborn, Susan, that seemed to be chess—at age three she stumbled on a chess set in a cabinet and bombarded her mother with questions about the strangely shaped pieces. The family found that Susan, ordinarily an exhaustingly active child, calmed down when she focused on the game. But was it coincidence, then, that his second and third daughters also would identify chess as their main passion? Or was it just the instrument to which their attention was turned with exceptional focus? The book also lacks many clues as to how László, in just a year, turned Susan into a competitor so fierce that while still only four she won a local championship that included students as old as ten. Soon after that, her parents took on the local Hungarian government, which did not approve of homeschooling, an essential component of László's plan, as it meant they could devote more hours of the day to the game than school would allow for.

László's equation for raising a genius, based on hours of reading about geniuses before he even had children of his own, boiled down, as he put it, to this: Genius = work + favorable circumstances. Those favorable circumstances included parents willing to commit to constant and early pedagogy—focused instruction for four to five hours a day, starting very young, in a particular, specialized domain. László was so convinced that he could make any child a chess genius that he boasted about it; at some point, a Dutch billionaire, Joop vanOosterom—a chess and billiards enthusiast who sponsored competitions in both—offered to help László, as Susan recollects in *Breaking Through*, "adopt three boys from a developing country and raise them exactly as they raised us." Susan said that László considered the offer seriously, but her mother refused. "She understood that life is not only about chess and that all the rest would fall on her lap," she said.

Their mother, Klara, may have also understood that her daughters were not so much a reflection of their father's pedagogical gifts as they were, as one neuroscientist theorized in *Psychology Today,* "a beautiful coincidence." That researcher, Ognjen Amidzic, born in Serbia, had gone so far as to move to the Soviet Union as a young man to pursue his dream of becoming a chess grand master. Amidzic—who could never achieve the level of grand master, despite his dedication and training—eventually turned his talents to neuroscience, trying to identify neurological differences between those players who were dedicated and well-trained amateurs and true grand-master chess talents. What he found suggested that the theories that David Epstein applied to the "sports gene" might also apply to chess players—that while having some innate talent wasn't enough to make someone a star, nor would motivation and superb training alone, without some innate gift, be enough to achieve outstanding results at the grand master level. He found, among the world's best chess players, a particular quality that he believed couldn't be taught: Chess grand masters, scans suggested, relied on parts of the brain associated with long-term memory and higher-level processing, possibly giving them access to a vast database of thousands of games they'd already played. (The average dedicated chess player, by contrast, relied on short-term memory. "Amateurs are overwriting things they've already learned," Amidzic told *Psychology Today.* "Can you imagine how frustrating that is!") Since then, the debate has continued over how much those kinds of memory skills at the highest levels reflect innate capacity or training. But in Amidzic's assessment, it was not that playing chess all those hours a day gave the Polgár daughters this unusual brain capacity, as he imagined it; rather, what struck him was that this particular gift might have gone unexploited, wholly unrecognized, had the girls' father not had the unusual drive to devote their very young lives to chess.

The idea of adopting a child to prove some theory about the power of their parenting would have been anathema to David and Micki Colfax, authors of the child-rearing guide *Homeschooling for Excellence*, who did adopt two sons. Progressive academics and activists, they raised, in total, four sons, right around the same time that the Polgárs were preparing their daughters for greatness. The Colfax parents were, for a brief moment in the eighties, celebrated on talk shows and in countless news articles, renowned as homeschoolers who managed, despite providing their children a highly unconventional education, to send three of their four sons to Harvard, a feat that generated considerable press. (Harvard is an extreme environment of its own that comes up a lot in their book, and one that poses its own chicken-egg questions: Were the school's admissions committees truly deft at picking future public leaders in their fields, or did the brand, privilege, and connections that come with Harvard hasten their ascent?)

Eventually, their oldest, Grant, became the director of public health for San Francisco, after serving in the Obama administration as its director of national AIDS policy. Drew Colfax received a law degree, then went back to Harvard for medical school, returning to Mendocino County to become an emergency room doctor; and Reed, the older of the two adopted children, is now a partner at a nationally recognized civil rights law firm, having dedicated himself to issues of fair housing. Garth, the younger of the two adopted children, works with the developmentally disabled.

David Colfax did not set out with deliberate ambitions for his children, as László Polgár so clearly did. He was devoted, instead, to his work as a sociology professor, researching the social structures and environments that limit or enable individuals' potential. His politics and academic writings were considered radical—so radical, in fact, that he was denied tenure at Washington University in St. Louis after being asked to account for his support of the Black Panthers and for critiques he'd made of his field as being too politically conservative. His difficulty in finding work was part of what led him and his wife, Micki, who had a master's degree from the London School of Economics, to make the choice to join the then-thriving "back to the land" movement. In 1972, they bought forty-seven undeveloped acres of land on a mountaintop in the redwoods of northern California, less interested in creating some alternative, ideal learning environment for their children than in finding a permanent place to live that they knew they could afford.

In their memoir, *Hard Times in Paradise,* David and Micki Colfax are unstinting in capturing how demanding, exhausting, and, at times, dangerous their work was, as they transformed wooded acres into a homestead and eventually a goat farm. "Living on nothing, an incredible drag, worrying about money all the time," Micki wrote in 1975 in her diary, which she excerpted in their book. "The trouble with being poor is that it takes up all your time." There were months of showers from sprinkling cans, doing laundry on a washboard, cooking on a camp stove in the rain. There were years of physically hauling water up to the house from the closest source, a project in which the boys all participated. Ingenuity and some degree of luck were a matter of survival (as when Grant made the wise choice to whack a charging steer with a plank of wood that he happened to be holding, rather than trying to outrun him).

They had a screen-free household for the same reason they had no telephone: because they had no electricity. The only conventional forms of entertainment available to their children were books, board games, and occasionally a radio that played old-time radio classics like *The Shadow.* The three oldest children all learned to read at about the same time, almost spontaneously, when the oldest was nine and the younger two were six and four. They turned to the *World Book Encyclopedia* when they needed to understand the workings of a carburetor on the family's malfunctioning generator or to learn if the redwoods that surrounded them could be found anywhere else. Using what the oldest son, Grant, gleaned from books about Native American people, the boys, on their own, cleared land and built a thousand-square-foot garden protected by a seven-foot-tall fence of poles and branches woven together. The boys built themselves a tree house in an oak grove, and they studied calculus and genetics and embryology as they got older. To supplement the money the family earned by selling wood, Grant started raising prize goats; the boys also slaughtered pigs for market. "I think the four of us knew more about the anatomy of an animal than anyone at Harvard," Drew told the *Associated Press* soon after arriving in Cambridge. "I just had a white rat dissection lab that was so boring. It was, 'Here's the stomach and here's the heart.'"

Their life was an ongoing exercise in on-the-spot education; judging from Micki's diary entries, it was also grueling, dangerous, and only sometimes exhilarating. They were trapped—as much by penury as by commitment—into continuing to live on rice and beans, dressing in Goodwill items, reading by kerosene lanterns. Their home was "a beautiful Siberia, but I'd leave it for a good teaching job," David Colfax told one San Francisco reporter in 1992, looking back on the isolation of their home life.

And yet, in 1988, after their three oldest sons received so much favorable attention, David and Micki published a book, *Homeschooling for Excellence,* as if their experience could provide a primer for others hoping to achieve similar results. "Drew, at seven, understood that the Pythagorean theorem was invaluable in squaring up his sheep shed foundation," they wrote, in one example of the kind of enrichment their lives necessitated. Micki credits the hard work required to simply survive as a factor that made the children cherish the rare time they had with books; Grant went from not reading at all to mastering reading quickly, at age nine, to satisfy that new fascination with Native Americans, and within a year, he was reading "college-level anthropology monographs." The Colfax family offers the would-be-successful homeschooler suggestions for the best math textbooks (Spectrum mathematics workbooks, still available) and writing guides (Strunk and White's *The Elements of Style* and William Zinsser's *On Writing*

Well). But of course, what made the boys' knowledge so deep, their competencies so powerful, was not particular textbooks but an upbringing that was idiosyncratic and exceptionally challenging. It could offer the typical parent few replicable paradigms, although I myself did draw from the Colfax family's story the reinforcement of one notion I believe to be true and that Julia Leonard's research bears out: Children are capable of so much more than we think—and they thrive when they are made to feel that their contributions are vital. In the case of the Colfax family, they did not have to find time to create or contrive those opportunities; the urgency was impossible to ignore.

The Colfax brothers grew up to lead meaningful professional lives of high purpose; they were also raised by parents who aspired to lives of high purpose. "If you want your kid to be a better human, the way to do it is to *be* a better human," the children's book author Mo Willems once said in an interview in *The New York Times*. (More easily said than done, for most of us.) The Colfax children's only regular role models had extraordinarily distinct values; they were also exceptionally well equipped to pass on those values, as they were not just brilliant individuals but professional educators (as were Jeannine Groff, Millicent Holifield, Ellis Marsalis, and Esther Wojcicki). After acquiring a master's degree from the London School of Economics, Micki Colfax became a high school teacher, a job she held before the family uprooted their lives and started over. The children were living in the hermetically sealed universe of what they read in books and what their parents modeled. Rarely are siblings' shared environment—which includes having the same parents—so pure and separate from the whole world of individual experiences that most kids encounter, a realm that includes jostling with other children and adults in their schools and neighborhoods. For most of their upbringing, the Colfax children attended no schools; they had no real neighborhood. They had their parents, their books, and their wits, which were exercised and taxed at least as many hours a day as the Polgár girls played chess.

Isolation, self-imposed, may have reinforced the parenting and sibling effects they had on one another—who was there to offer any other kind of messaging or distraction? As I read about the Colfax family, I found myself thinking about Serena and Venus Williams—how their mutual reinforcement might have been all the more powerful because they, too, were somewhat isolated. They had each other and their three older half sisters, but, growing up, they all had few friends, Isha Price, an older sister to Serena and Venus, told me for a profile of Serena Williams that I wrote in 2007 for *The New York Times*. The family—the entire family—had a lights-out ten o'clock bedtime, which Isha told me she didn't realize was unusual until she got to college at age eighteen. The girls' room held only four beds for five girls—so the girls took turns making room for Serena to squeeze in. Like the Murguía twins, like the Colfax family, like the Brontë sisters, they were all so close, emotionally and physically.

David Colfax, in his professional discontent, also falls squarely into the category of the thwarted parent. "If we hadn't moved up here, we'd be middle-class white kids living in the suburbs with a professor for a father, getting As and playing varsity sports," said his son Drew to a reporter in 1986. "It definitely worked out the best for me, but I feel bad that my dad isn't able to do what he chose to do." At the same time, David and Micki Colfax had extraordinary agency: They made the successful Herculean effort, over a decade, to transform, with their own hands, all those acres of uncleared land into a functioning home and a viable farm, despite no previous experience in building or agriculture. They were extraordinary; they were overcomers. They, too, would have likely identified with the sentiments that Patrick Brontë expressed to the biographer Elizabeth Gaskell (true also of László Polgár, and certainly of Richard Williams, Venus and Serena's father): "If I had been numbered among the calm, sedate, concentric men of the world, I should not have been as I am now. And I should in all probability never have had such children as mine have been."

WHAT MADE LÁSZLÓ POLGÁR and the Colfax parents so unusual is that both created dramatically unusual environments for their children: László by choice; the Colfax parents, as a by-product of necessity and their own desire for a different way of life. What helped make the Wojcicki children's upbringing so exceptional was their unusual environment as well: Palo Alto in the 1970s and 1980s. Their parents did not create that town's culture, but they chose it.

To create a culturally rich environment, Esther Wojcicki did not need to labor to lure educated, cultured people to her children's side, as Ying did with her restaurant, with free food, praise, and charm. Esther merely had to open the door and let her children go outside. The neighborhood they lived in was known as the "faculty ghetto"; they swam alongside the children of Stanford faculty at the scrappy local swim club, competed against them in (honors) classrooms.

The next-door neighbor and surrogate grandfather whose lemons the girls picked was George Dantzig, a mathematician who developed tools that revolutionized the way governments and businesses made complex decisions. The girls would run to his home for Wrigley's gum and York Peppermint Patties and end up learning a chess move or competing to answer an impromptu math problem. The girls' friend, Karen Levy, now a professor of environmental and occupational health sciences at the University of Washington, lived next door to Amos Tversky, the behavioral economist who would have been the co-beneficiary of the 2022 Nobel Prize, along with Daniel Kahneman, had Tversky not died in 1996. Karen's father was head of oncology at Stanford for two decades, a man whom many were looking to for the cure for cancer, just as they were counting on Stan Wojcicki to broach the next frontier in particle physics. Knowledge was everywhere, as accessible as the lemons they picked from the trees, as free and as indulgently shared. Anne and Karen only had to get stuck on a math problem to have an excuse to call—at eight, nine, ten o'clock—the school's math hotline, where their favorite teacher would talk to them for hours about theorems, the concept of zero, the meaning of life. And school and home were not the wholly separate worlds they are for so many children; by the time Anne was in eighth grade, Esther was on her way to becoming a legendary journalism teacher at a high school in the same school district.

What is striking to Anne is not that she and her sisters all went on to break into male-dominated fields, but that they were far from alone. The girls were friendly with two other families of three girls; in both families, the girls went on to have major careers in law, in business, in science. One friend was on the board of eToys; Karen Levy also runs her own research group at the University of Washington. "Some of what people do is pretty esoteric, and they are all really passionate," said Anne.

Susan Wojcicki has shared the same impression of their community: "We had all these amazing people around us and they really cared about doing something meaningful for the world," she told *Fast Company* in 2014. Their goal wasn't to become famous or to make money; it was to do something meaningful for the world, because they had a passion, they found something interesting, and they cared about it. "I mean, it could be ants or it could be math or it could be earthquakes or classical Latin literature," Susan said.

Anne Wojcicki and Karen Levy had another friend with whom they were so close, it was as if the three of them each had three mothers. Even as an adult, when Anne visited the Levys, she called Karen's parents Ima and Abba, Hebrew for "Mother" and "Father." The Wojcicki sisters were raised by fierce Esther, but also by a woman, the mother of a close friend, who was a profoundly empathetic therapist, and by Karen's mother, Shoshana Levy, a prominent biochemist and oncology professor at Stanford who showed more of a tough-love, no-excuses approach.

In 1998, a force entered the Wojcickis' lives that probably had at least as much of an impact on their roles as innovators as even Esther: Sergey Brin, a computer science PhD student at Stanford launching a new business. Brin, along with Larry Page, had mentioned to a friend of Susan Wojcicki that they were in need of a start-up space. Susan, already in possession of two degrees (one in business administration), was working at Intel and had just purchased her first home. She decided to rent out her garage to the two young men. "A year from now, people will know about our search engine," they promised her. Susan remembered thinking, *If you're so important, why are you renting my garage?*

Over many take-out dinners, she got to know Brin and Page, and she eventually quit her job to join them as their company grew. She was employee number sixteen at Google.

At the same time, Google might not have been Google if it were not for Susan, who first persuaded universities to partner with the company by including its search tool, free of charge, on their websites; Susan was also the person in charge of finding a way for the company to make money from its powerful and popular search tool. Susan helped develop AdSense, a program that would serve as an invisible middleman, dropping advantageous advertising content onto websites without effort on the part of whoever ran those sites. Later, she made a pitch for Google to purchase YouTube for $1.65 billion, an investment that generated upward of $40 billion in revenue in 2023.

While Susan was building a name for herself at Google, Anne, who had graduated from Yale, was working in biotech investment. Susan introduced Anne to Brin. She and Brin—both irrepressibly confident and athletic—soon started dating.

In the early years of the new millennium, Anne's boyfriend grew a business from a garage-based start-up to a billion-dollar valuation. Her own sister had thrived in a technology business, demonstrating the rewards of taking a risk. "I remember when my sister first launched

AdSense, she was like, 'I don't know. Let's see how this goes,'" Anne told one interviewer. Susan didn't have to be sure of herself to take a chance. She just had to have a hunch and the confidence to act on it.

Anne herself had a hunch: She thought it would be important to find a way to provide solutions outside of a health-care system that was not rewarded for providing preventive care. And as someone who was trained to question authority, she was not afraid of putting information directly in the hands of consumers, rather than relying on doctors as mediators. Why not think big by providing consumers access to their own DNA testing? Not only would patients, armed with knowledge of their genetic risks, be likely motivated to make lifestyle changes—becoming more physically active, eating a healthier diet— but the business, if it grew large enough, would have a network effect: With hundreds of thousands of sequenced genomes, scientists could far more quickly diagnose and come to understand the underpinnings of serious illnesses. With the equivalent of a $2.6 million loan from Brin—whom she later married, then divorced—Anne started 23andMe, a business that was worth $6 billion at its peak.

The girls were deeply plugged into the resources of their hometown and came of age just as that hometown—their backyard—was becoming a locus of innovation and commerce to rival any in history. Susan's career was well on its way to prominence when Anne entered the game; but would Anne have had the resources, the funding, the inspiration, without having an older sister who found her footing at a start-up before her? When sisters both acquire influence at such high levels, especially in a field that is male-dominated, it is unlikely to be a coincidence—so often they do it with one another's help.

When I spoke to Karen Levy, she, like Anne, brought up the many successful three-sister families they knew from Palo Alto. "If any of us had brothers, I wonder if we'd have been as encouraged," Karen said. "There was no option of giving more encouragement to a boy—so the three girls were told they could do what they wanted to do." Perhaps that dynamic was also at play in the home of the aforementioned Sullivan family, which had four daughters (two of whom became CEOs of Fortune 500 companies)—and no sons.

A Danish microeconomist named Anne Brenøe wondered about the same phenomenon, ultimately concluding, as the title of her 2022 paper indicates, "Brothers Increase Women's Gender Conformity." Relying on a massive database of Danish administrative data, she found that women who had younger brothers, as opposed to younger sisters, were more likely to choose careers dominated by women—which tend not to be in STEM fields and tend not to be as well-compensated as traditionally male fields. Sisters with a younger brother even earned less after having a child than did sisters who happened to grow up with a younger sister. "The random event of the arrival of a brother to the family instead of a sister changes these women's socialization within the family to such an extent that they are put on a different career path early on," she wrote, having studied the daughters' educational choices even before they left the home. As for Brenøe herself, she has a younger brother—which doesn't necessarily contradict her research, she said: "Who knows? Maybe if I had a younger sister, I'd be a physicist, not an economist!"

In the Brontë sisters' childhood home, professional ambition was largely prioritized for Branwell. But the sisters reflected back at each other a sense of possibility and, more practically, served as springboards for each other's ideas: Martha Brown, their former housekeeper, recalled in an interview for Gaskell's biography that, when they were working on their novels, they often stayed up past 11 P.M., walking around and around the kitchen table, discussing their work and talking through their plans.

The ritual was so much a part of Charlotte's life that she continued the habit of walking around the table even after her two sisters died—Emily, at age thirty, a year after publishing *Wuthering Heights,* and Anne, at age twenty-nine, around five months later. "My heart aches to hear Miss Brontë walking, walking on alone," Brown wrote to a family friend.

The three women wrote under pseudonyms, and as close as they had once been to Branwell—Charlotte once wrote that she and her siblings were all minds "cast in the same mould"—Charlotte believed that he died without knowing their great secret. "My unhappy brother never knew what his sisters had done in literature," Charlotte wrote to her publisher shortly after his death. "He was not aware that they had ever published a line; we could not tell him of our efforts for fear of causing him too deep a pang of remorse for his own time misspent, and talents misapplied. Now he will never know. I cannot dwell longer on the subject at present; it is too painful."

They might have kept their secret for more practical reasons, as well: the concern that he would ask them for money, if he knew they were starting to earn it, the critic Claire Harman suggested to me, or that he would in some other way impinge on the progress they were making to move their lives forward as artists and earners. Their decision to keep from him the astonishing news of their success speaks to how fiercely they protected that newly formed, all-female collaborative.

FRANCIS GALTON, the author of *Hereditary Genius*, published in 1869, perfunctorily included the Brontë sisters in that catalog of family talent. He shoved them into an appendix—"Appendix to Literary Men"—and was sloppy with the facts: He got Charlotte's age, at the time she published *Jane Eyre*, wrong by nine years and attributed her death to consumption (she tragically died, while pregnant, of complications). He proclaimed Branwell "perhaps the greatest natural genius among them all."

Were he to have studied the Wojcickis, he might have looked back into their illustrative family history, which started long before Stan and Esther met in a dorm at Berkeley. Stan's mother, Janina

Wójcicka, a Polish immigrant to American via Sweden, was an esteemed librarian of Polish works at the Library of Congress, a woman for whom the collection of books was a matter of national significance and a professional passion. But that was only her last career. Before that she had been an academic, acquiring her PhD in Poland, studying Western cultural influences on royal leadership in Germany in the fourteenth century. "I never knew another woman of her generation who had a PhD," said Esther, who considered her mother-in-law extraordinary.

Janina's husband, Franciszek Wójcicki, started out from a family of farmers, but became close to a more affluent family who all but adopted him when they lost their own son. He became an appellate judge, but never forgot his roots, championing the rights of the working class. Following the Nazi invasion of Poland, he was evacuated to serve with the Polish government in exile and was a representative for the Polish People's Party in that exiled government. In hearing Anne Wojcicki talk about her work, I was struck by how much she has in common with her grandparents. "I wanted to have a company that was, frankly, really rebellious," she said in one interview. "It had a Robin Hood mentality. How to build the future? You can't change the system from within, so we've gotta do something more radical from outside."

The author Galton played a large and terrible part in the Wojcicki sisters' grandparents' lives and those of so many others. Galton has been recognized for his (crude) breakthroughs in twin studies, which eventually compared identical and fraternal twins to try to glean the power of heredity, operating off an insight that became foundational for behavioral genetics. But he is also infamous for the way that his intellectual curiosity meshed with the ultimate elitist perspective on the world. In coining the term *eugenics*—from the Greek "good of birth"—he helped unleash some of the most vile historical consequences in modern history, in Nazi Germany and beyond, consequences that he could never have imagined. In Germany, his science was used to justify a theory of Aryan supremacy and antisemitism, the same insidious bias that drove both Esther's parents, Jews from Ukraine and Russia, to the United States in search of more opportunity and less oppression.

With its data, 23andMe helped make it possible for geneticists to complete some of the earliest efforts to find associations between thousands of genetic variants and educational attainment. Researchers' intent in seeking to identify those associations is to help them do better research, by controlling for genetic factors in large-scale studies on, say, the beneficial effects of attending preschool. But in a paper published in *The New England Journal of Medicine* in 2021, many genetic researchers, including some who worked on those genome-wide association studies on educational attainment, also call for "a society-wide conversation" about the way that this new form of research might be used irresponsibly. They caution against companies, specifically, that offer to test parents' embryos for reasons that go beyond medical risks—that offer parents the promise of choosing from among their embryos the one that has the greatest chance of completing college, for example. To begin with, the paper argues, various factors make it unlikely that embryo-testing companies will be able to provide enough information for parents to make well-informed choices or fully appreciate how limited the differences actually would be. Beyond that, they point out, simply offering certain choices is problematic in and of itself. Were those companies to offer parents embryo selection for a certain skin pigment or hair color, they wrote, that might "reinforce racist conceptions of biologic superiority by signaling, either explicitly or implicitly, that certain traits carry value or stigma, possibly amplifying racial prejudice and discrimination." The paper flags the concern that Galton's notion of the "betterment of the species" might emerge in a different form, in different circles.

In recent years, 23andMe has faced legal action and criticism over privacy breaches; it has plummeted from a business valued at $6 billion at its peak to one that is a penny stock, struggling to maintain its place on the Nasdaq, and as of the fall of 2024, Anne's plans to take the business private have yet to be realized. Anne spoke, at the start of her business, of her intention to give people better means of providing their own preventive care: If you know, as her ex-husband, Sergey Brin, found out, that your genetic profile puts you at a higher risk of Parkinson's disease than the average person, then maybe you take precautions that some scientists hope can prevent it. But knowing those risks did not turn out to change people's behavior, as the advice they received was usually no different from what they'd been hearing all their lives—exercise more, eat healthier. The company was also limited by a saturation of the market—it was not positioned to attract repeat customers.

The tests' findings were in other ways life-changing for many families, who discovered that their relatives were not who they'd always assumed; but the revolution in health-care treatments that genetic testing was meant to galvanize has been slower to arrive—certainly too slow for companies like 23andMe to effectively monetize. Like her father, whose dreams for the Superconducting Super Collider aimed so high it was unachievable, Anne was, history might find, ahead of her time: At some point in the future, the medical field and individuals may be able to choose to act on genetic testing in more targeted ways. In the meantime, I would argue, the popularity of her product, whatever its profit margin, has helpfully normalized a basic understanding of genetics among mass audiences—while also normalizing genetic testing in general, including, quite possibly, the kind that other companies might offer for embryos, which many have pointed out could do such harm.

Susan Wojcicki also helped create, with Google, a library of knowledge that is unfathomably more vast than any her grandmother could have imagined. If 23andMe knows some of who we are through our genes, Google knows so much about who we are through our clicks.

By reflecting ourselves back to us—through advertisements, through YouTube video suggestions generated by an algorithm—it creates an environment that may further reinforce who we intrinsically are inclined to be, without even requiring us to leave our desks.

Perhaps Esther Wojcicki really did raise girls the likes of which the world has never seen. They built highly technical datasets of superknowledge that are changing the human experience, with repercussions still unknown. She helped unleash their talents; what they have helped unleash suggests the vast yawning chasm between our intentions and the unpredictable ways the world acts on them.

TWELVE

Better, Better, Best

Although it's fascinating to try to tease apart the elements of nurture and nature that, in some combination, might ease some families' path toward success, it's also practical to consider the role of luck, which showers down unevenly not just on siblings but also on entire families. Just by the law of probability, some families will inevitably be outliers, the beneficiaries of more than their fair share of happy happenstance. "Every once in a while, in a very rare while, if you pull the slot enough times, it is going to come up all cherries," said Dalton Conley, the Princeton sociologist, when I first called him to ask his thoughts on the topic of families with numerous high-achieving siblings.

For most of his career, Conley has studied the structural and economic explanations that drive most individuals' life outcomes. His body of work, however, betrays a particular interest in research on the nature of family and how it shapes people's lives, with insights that changed over time as new findings emerged, which is why I so often found myself turning to him for his perspectives. In *The Pecking Order*, published in 2004, he explored the ways that common fluctuations in family financial stability so often lead to income inequality between adult siblings, because they had different opportunities along the way. A decade later, he turned his own informal ongoing experiment in child-rearing into *Parentology: Everything You Wanted to Know About the Science of Raising Children but Were Too Exhausted to Ask*. Married at the time to an artist and engineer named Natalie Jeremijenko, Conley set out, as he writes in *Parentology*, to test various theories of parenting, some standard, some of the couple's own devising—and to encourage other parents to do the same. "For there's simply no one-formula-fits-all to raising successful, compassionate kids in today's impossibly complex and radically overstimulating world," he wrote.

The experimentation started, for each of his children, at birth. Conley had read research that found that children with unusual names had higher levels of self-control; the theory was that children with unusual names were forced to be calm in the face of blowback every time they met someone new.

So their children's names would not be conventional. Conley and his spouse decided that their daughter would not be subject to assumptions people made about her based on a choice her parents made before they knew her. They would call her only E, and when she was old enough, she could decide on a name herself. Of course, not giving a child a name will also allow people to make assumptions about that child, based on the assumptions they make about that child's parents: They think of themselves as inventive; they have high hopes for their child's uniqueness and have been sending their child that message from birth.

Conley and Jeremijenko took a different route with their son, born two years later. They named him Yo, an ethnically ambiguous name with which Conley hoped his son would challenge societal assumptions.

Starting in the first years of his children's lives, Conley had taken note of a body of literature, suggesting that a child's exposure to complex syntax and language—not some innate ability—was what determined their later literacy. As Conley wrote in an article in *Time,* pegged to the release of *Parentology,* he did not feel equipped to keep up a steady stream of sophisticated discourse with his toddlers; instead, he wrote, "I decided the best thing was to read to my kids constantly." He not only read to his children but also chose books significantly above their grade level, frequently stopping to define words or offer alternatives ("like a human thesaurus or a backup singer," he wrote), indulging, when the kids got old enough, in side conversations about everything from organ transplants to the difference between slang and dialects. "So while E and Yo were both behind their peers in reading in first grade," he wrote in his essay in *Time,* drawing a cause-and-effect conclusion that would likely not withstand peer review, "by fourth grade, they had the best scores in their respective classes."

Conley's book reveals a kind of maximalist approach to parenting that feels all the more excessive, given the advantages they already had as offspring of Conley, a double-PhD Princeton professor, and Jeremijenko, an artist-engineer and, for a time, an NYU professor whom *Fast Company* described as "one of the most influential women in technology." They had at their disposal the benefit of good schools, safe housing, and a hometown rich in cultural resources: New York City.

Even at the time he wrote the book, Conley was, and still is, like his fellow sociologist David Colfax, a strong believer in the mighty power of the environment to shape an individual's destiny. But he was already starting to rethink a deep skepticism he previously held about the field of behavioral genetics, whose studies generally suggest that genes also play a significant role in influencing aspects of individuals such as their personalities and mental health. He thought there were likely flaws in the methodologies of twin studies, which were, for many years, the basis of the field's findings, and which allow researchers to make inferences about how much genes and the environment each account for why people differ from one another for a given trait. The studies' premise works like this: If, in studying thousands of twin pairs, researchers observed that identical twins (who are made from the same genes) were more similar to each other in, say, conscientiousness than were fraternal twins (who only share about 50 percent of the same genes), then they could infer that genes played at least some role in influencing that particular personality trait. The thousands of twin studies that have been conducted over the years—studying coffee consumption, gun ownership, television enjoyment, to name just a few—suggest the near-ubiquity of at least some genetic influence on most aspects of our identity. (This is not to say there is a gene for, say, coffee consumption—rather that hundreds, or maybe even thousands, of genetic variants work collectively, in interactions with the environment, to yield a more likely result: a taste for coffee.) On average, one major meta-analysis found, the breakdown of genetic and environmental influences in accounting for differences across individuals for some seventeen thousand traits, including medical, physical, and psychosocial ones, was almost evenly split.

Conley devotes several pages of *The Pecking Order* to implicit critiques of twin studies, which he (along with many others) believed operated from a problematic premise: Even if identical twins were much more similar for a given trait than fraternal twins, perhaps the genetic component was being overestimated—maybe the world responded to identical twins in unusual ways, treating them like a unit, for example, or mistaking them for each other; certainly, they tend to spend more time with each other than fraternal twins do. Some identical twins dress alike; fraternal twins, by contrast, might deliberately differentiate, not as a function of their genetic destiny, but because of the psychological impulse to individuate. For many reasons that had nothing to do with their genetic overlap, identical twins might show more similar outcomes than fraternal ones.

After he published *The Pecking Order*, it dawned on Conley that he could definitively prove the flaws in the methodology of twin studies. By then genetic testing had made it possible to know with certainty whether twins were monozygotic (identical, split from one zygote), or dizygotic (fraternal, originating with two separate eggs fertilized by two different sperm). Conducting a genetic analysis of twins in a large database, he found that a surprising number of identical twins had been mistaken for fraternal twins (conditions in utero is one way to explain how identical twins might actually look different enough that they'd be mistaken for fraternal ones). Now he could do new studies, but this time, with the correct classifications. Conley expected that in his new analysis, the identical twins whom the world (mistakenly) perceived to be fraternal—distinct and different—would be less similar in their grades, for example, than identical twins recognized as identical. These "fraternally raised" identical twins would be found to be not quite so similar after all, which would make the gaps between fraternal twins and identical twins for trait similarity less dramatic.

He did the analyses for several traits and was surprised by the results each time. The usual twin-study breakdowns held up, and for some traits and conditions (for example, ADHD), the results even showed more hereditary influence once the misclassifications were cleared up. "I had come into this field thinking, I'm going to show the geneticists wrong, social science is going to rule," he said in a TEDx talk in 2015. "And here I was, and I couldn't knock a single estimate of the genetic effects down."

By the time Conley published his findings in 2013, he had already enrolled in his second PhD program, an unusual choice for someone that far along in his academic career. The field of study was even more unusual for a sociology professor at that moment in time: biology, with an emphasis on genomics. He had become increasingly curious about the results that would emerge if sociologists cross-referenced their own work with huge, burgeoning databases providing genetic information that could be tied to important measures, such as educational attainment. He still expected environmental effects to be powerful; but wouldn't scientists be better able to measure which educational or family interventions worked—and for whom—if they could control for genetics? "I don't know why other people in social sciences don't feel like there's a huge gauntlet thrown down by all this research," Conley told me. "If you want to best explain the developing behavior, you need to consider both nature and nurture."

Conley was in his biology PhD program and studying genomics as he was putting the finishing touches on *Parentology*, a book that could be read as presenting parents as capable programmers of their children, experts who need only the right manual to achieve their desired results. His own research had, if anything, only bolstered a finding that undermined the sales pitch of the book. When twin studies break down the significance of twins' shared environment, such as being raised by the same parents, they do find that it has a profound effect on how much schooling those children receive as well as their religious affiliation, truly meaningful factors in their lives. But for most traits, they usually find the influence of the shared environment dwarfed by the different environmental circumstances that children encounter, the vast number of idiosyncratic influences that come to them over the course of their lives: the demanding geology teacher; the broken ankle at a Coachella concert; the especially sunny dorm room; the friend who passed along a job listing; the millions of unique, often random, experiences that bear down on us, interact with whatever genetic leanings we do have, and tweak our destinies in an ongoing, endlessly iterative way. Identical twins have the same parents and are raised in the same time period; the same is true of fraternal twins. But if parenting were a massively powerful influence on personality, then fraternal twins would presumably be more similar to each other for various traits than they generally turn out to be.

Conley presents this evolution of his thinking in *Parentology*, while stopping short of throwing up his hands at the futility of thinking that his elaborate parenting efforts were all for naught. He argues that there are many clear ways parents can shape their children's tendencies—for example, he points out, since voters tend to keep going once you induce them to vote once, parents can have a lasting impact on their children's civic engagement just by ensuring that they get to the polls. And he would never consider all the energy he exerted to help his children a waste—all his efforts on his kids' behalf, all that reading to them, all that offering of enrichment. If it did anything, he wrote, it showed how much he cared: "Scientific parenting is my ode of parental love," he wrote.

On the Thanksgiving when I first met Dalton Conley's two children—I was curious to hear from his research subjects—they were back home on break from college. Conley and Jeremijenko had divorced by then, and as he and his children sat on a couch in Conley's loft apartment in Lower Manhattan, their stepmother, an astrophysicist named Tea Temim, then four months pregnant, was busy preparing and roasting tray after tray of vegetables. Sound echoed in the spacious loft, with its concrete floor, and as Tea worked, she could half listen to Yo and E reflecting back on their lives as experiments, the results of which their father had awaited anxiously all those years. I was asking them about their upbringing, now that they were old enough to look back: How had they felt about being test subjects? Did they think their father had rigged a way of making them somehow better versions of who they otherwise would have become?

Conley's daughter, E, was a junior at Princeton, and her brother, Yo, a sophomore at the University of California at San Diego. E wore jeans and matte gold boots, city-kid dress for a lazy Thanksgiving morning. She smiled encouragingly across the coffee table, looking perched and ready; her brother, sitting beside her, was laid-back and vaguely anticipatory, one leg up on the table, his arms stretched out as if in a perpetual yawn. Yo, who as a child was a mediocre math student—he would never show his work on tests and in homework, especially since it came together so quickly in his head—was now studying math and economics, preparing to compete, in a few weeks, in the William Lowell Putnam Mathematical Competition, one of the most rigorous in the country. E would be graduating the following spring and wanted to be an actress or maybe a writer.

E never did decide to choose her own name, having grown accustomed to the initial and even liking it. "Names should be something that feel special," she told *The New York Post* once, while being interviewed on the subject. "Every day, I would take the E train to high school. If I had a bad day, I would pretend the train was just for me."

Conley, sitting beside his two children, had errands to do before the Thanksgiving dinner; he looked a little restless and uncomfortable. *Parentology,* he had come to feel, was a project he regretted, mostly because it was intended to be more humorous than his editors understood, but also because his children had been works in progress, too fledgling to be memorialized any one way—too young, really, to have given consent. Reading it once his children were older, I found the book endearing in its dramatic irony—its bold proclaiming of certainties that would give way as life unfolded. Because of the risk of brain damage, Conley's own child, he swore in *Parentology,* would never play football; in fact, Yo not only played football, but was a serious enough player that his father cheered him on at scouting camps (though only ones in which the athletes competed for Ivy League teams).

Over time, Conley was increasingly growing interested in the way that genes and environment function like "two sides of a Mobius strip," as he put it, factors that can't be seen as isolated and separate forces, a theme he explored for his most recent book, *The Social Genome.* Genes might nudge someone toward certain behaviors, which then elicit a certain response from the world, so that one's environment is not random but tailored. A child might be genetically nudged toward becoming a reader—but if that child's parents refuse to read to that child, the effect of "nurture" in that home actually interferes with the way those genes are ultimately expressed.

"All genetic effects are contingent on environmental context," he said to his children, talking through his thinking with E and Yo that morning. Food scarcity in the 1930s suppressed the likelihood of obesity; only in more recent years, with the widespread availability of high-fat diets, his research and others' have found, can the differences in people's genetic propensity toward obesity even be expressed more widely. Now that higher education was more available to women, scientists could better measure genetic differences among them. "For women in the first half of the twentieth century to go to school, you had to be a very special person," said Conley.

"I have a question," said E. "What did you mean the women who went to college earlier were more special?" Conley explained that women had to be very unusual, both in personality and cognitive ability, to push against the norms of the time. E jumped in with her response: "The women who went to college in that time—that was a huge product of racial and economic privilege." Her father readily agreed.

"I'm going to use the bathroom," said Yo, who could see where this was going.

"I think that it is important when making offhand comments to acknowledge the larger social forces that existed, right?" E said. "Like when you just refer to someone as special, that implies intelligence."

Perhaps this kind of debate was common between Conley and E?

"I'll let her answer that," said Conley, who was starting to look as if he regretted the interview as much as he did writing *Parentology*. E did not believe in standard measures of intelligence or "anything like that," she said. "For example, being at a place like Princeton—I think it's barely anything about quote, unquote natural intelligence. It's about how much cultural and economic capital you have."

"Can I go now?" her father asked and excused himself to head to Whole Foods.

E and Yo, in Conley's absence, seemed happy to answer any questions about their upbringing. "I like interviews," said Yo, when he'd returned from the bathroom; he had done several of them as a thirteen-year-old, when his father published the book. E felt similarly: "I like meeting new people," she said.

Although they were clearly different in temperament, they agreed on one thing: That their father's creative parenting efforts were not designed exclusively for fun, but to help ensure that they excelled.

"The family motto was 'Better, better, best!'" said E. It was something they heard from their dad.

"It wasn't, like, emblazoned on a crest," said Yo. But it was definitely something they said and heard a lot—although maybe sort of a joking motto? "'Better, better, best!'" E said again, this time more quickly. The line had been adapted from a centuries-old homily: "Good, better, best. Never let it rest. 'Til your good is better and your better is best."

Yo remembered a feeling, as a child, that his parents were "grubbing around—seeing what we excelled at and what we could do well at." Yo was given piano lessons, voice lessons, guitar lessons, and chess lessons, whether he wanted them or not. And his father was intent on making math lovers out of them. In exchange for doing an extra twenty minutes of math beyond her schoolwork, E, as a young grade-schooler who loved sweets, could earn one candy peach ring. Yo, by bargaining, parlayed that into twenty minutes of *Club Penguin* on the computer. "It was a lot of incentives and rewards," he said. "Kind of like a dog trainer." It did not have the intended effect of making E, at least, love math; but it did have the effect, both said, of turning Yo into an expert negotiator. Eventually, he bargained his way into getting time and a half for every extra minute of math.

"Now he really is anti-rewards for things," said E of her father. "He's afraid it took away our inner motivation." Plus, he seemed to think he had created something of a haggling monster in Yo.

It pained her father, E knew, that she was putting off taking a major math requirement at Princeton: He knew she had the aptitude and was convinced that she had talked herself into believing that just because she preferred English and history, that meant she was bad at math. "I started out with equal aptitude for math as Yo," she acknowledged. Yo made an indeterminate noise of dissent. E turned and stared at him. Whenever her brother said something—or made a loaded noise—that irked her, she merely widened her already wide eyes even wider.

Their father was undertaking research that relied on polygenic scores—(still crude and scientifically controversial) estimates of an individual's genetically informed likelihood of having a given trait. Using data with those scores, he was trying to tease out whether families pegged their children somewhat arbitrarily, or in reaction to one another, into niches. He wondered about a student who had strong aptitude for math but was an even more outstanding athlete, for example. If he had a brother who was not as strong a math student but was extremely, noticeably unathletic, would the family come to believe that the second brother was, in fact, the better math student?

He was also interested in how, for a younger sibling, having a sister who had a high polygenic score for educational attainment might matter. While looking at actual educational-attainment results for research he has yet to publish, he was finding that a fraternal twin sister's high polygenic score for educational attainment seemed to improve educational outcomes for her female twin sibling (in males there was no such effect). Maybe she was enriching the environments around her more attentively than brothers typically did; or maybe she was evoking from her parents some sort of enrichment—asking for classes, or attention, or lively dinner conversation—that then carried over to her twin. (Research has shown that of all the sibling pairings, sisters are generally the closest.)

Mostly, Yo seemed to ignore his sister growing up; she used him as research, her first anthropological subject. She would sit in her room and take notes on the few things she could overhear her brother and his friends saying to one another amid long silences while playing video games. "Bro, I can bench-lift that frown into a smile" was her favorite. Yo rolled his eyes. "It was said as a joke."

Back to the topic at hand: their upbringing. "I enjoyed the competitiveness," said Yo. "I think it made me into a very competitive person." E felt that the energy was always less about beating someone and more about doing your own very best. "My brother is very competitive," she said. "I try not to be." Yo protested: "She says she is not.

But she is, too. She's closeted competitive." Again, E opened her eyes very wide and this time stared forward. "I'm sorry," she said. "That's not true." Where was the data? The evidence?

"Of course, if you work hard, you have some internalized ambition," she said. "But that is something I try to fight against. Because I think that ambition is just shame for your present self. And there is a lot of shame in all the academic settings I have been in, and I don't want to tap into any of those ideas. My freshman year at Princeton, I felt it sharply all around me."

Whatever their upbringing—the same twenty-minute math sprints, the same exposure to chess classes and singing classes, the same family motto—the two seemed no more similar than any two college students who lived in their building would likely be. As teenagers, Yo was always sneaking out to go to parties; E swore off ever tasting alcohol or trying drugs, even to the point that her own parents suggested that maybe she shouldn't rule out those things quite so strictly.

Now E and Yo were both grappling with the same issue: their desire to make choices their father did not fully support. He was still trying to optimize them. How could he help her become an actress? He did not know any agents or managers or even how auditions worked. And Yo did not seem interested in what his father wanted him to do, which was pick up and move to some new city to get a master's degree in data science; Yo just wanted to combine his current studies with a minor in data science. Even without the master's degree, he thought he might find demand for that skill among tech businesses.

"My dad's concern is that I will sell my soul and join Wall Street, which I don't want to do," he said. "His nightmare is that someone is going to offer me a million dollars and I won't be able to say no. That is not the worst fear to have." Conley wanted his son to follow the route he'd followed: getting his PhD and doing research at a university. But Yo wanted to take a different path at a tech firm. "I am trying to use economics and data science together to work on artificial intelligence," he said. At this, E looked sympathetically toward Conley, who by then was back from his stop at the grocery store. "I don't think he gets that," she said, referring to their father. This conversation was happening years before ChatGPT would arrive on the scene. She understood the possibilities already; she was of Yo's generation. His father's ambitions were not Yo's, for reasons not just of personality but of timing.

Years would pass and reveal, in this particular family, the effects of parental pressure and the power of a dream: Yo would graduate with a joint degree in math and economics and data science from UCSD and then acquire, after more pressure from his father, a master of science degree in data science at Columbia University, ultimately landing a job working for the NFL. (His father would still have preferred that he went on to get his PhD, but had also come around to thinking a high-paying job with a hedge fund could be all right.) E would be accepted to the prestigious graduate school for creative writing at the University of Iowa, where she would win a financial prize substantial enough to live on for a year after she graduated. She would move in with her grandmother in Lower Manhattan, providing back rubs in lieu of rent, excruciatingly aware of the privilege that allowed her to pursue her creative life.

And Conley would continue to worry about his children Yo and E, even though they were young adults. The endeavor of raising children in a way that will maximize their potential, that will prime them to succeed, is problematic for many reasons, including the mystery of knowing when the raising actually ends—and the mystery of determining how one knows a child's, or a young adult's, or an adult's, or a middle-aged person's full potential has been met. To the extent that parenting, as we think of it—this powerful, influential, determinative guiding force—never really starts, it also never really ends.

On that earlier Thanksgiving, Conley had rejoined the conversation with a look on his face that seemed to say: *Should I be worried about what I missed?* If anyone had made fun of him for overparenting his children, he'd have to prepare himself for more of it to come: Already he was worrying about how to keep his future child away from iPads. When Tea flew for work during her pregnancy, Conley insisted that she cover herself with a medical-grade X-ray shield. She had another blanket for lower-frequency magnetic fields and wore a mask (this was before Covid made them prevalent) to protect herself from possible viruses on the plane. She found a hair salon that used organic dye, but when Conley called to ask follow-up questions, he deemed the safety precautions insufficient for a developing fetus.

"I thought I would be more relaxed this time around," Conley said of his own parenting. But apparently not. How could he be? The world was only getting more dangerous, more competitive as it became more globally connected. He believed that were he to have applied to graduate school more recently, he never would have been accepted with the grades he had at the time.

At any rate, he had no regrets about the efforts he had made the first time around to make sure his children developed to their maximum potential, whether or not he believed that the science backed him up. It could not hurt to read a lot to your kids, just as it could not hurt to feed them healthy food. Think of it like Pascal's wager, he later wrote in an email, referring to the philosophical proposition articulated in the seventeenth century by the mathematician and probability theorist Blaise Pascal: It was kind of like the religious, philosophical notion that in deciding whether or not to believe in God, humans bet with their lives, with the potential downside of making the wrong call only substantive on the side of disbelief. Maybe better to believe? What parents do to enhance their children's learning might not help as much as we like, he said. "But we could be wrong," he continued, "and it could matter a lot. Since it can't hurt, why not emphasize it?" Besides, as a professor, his values were books, not movies or video games. That was what he wanted in his home.

He needn't have worried or felt defensive about any of it. In fact, while he had been at the grocery store, E had basically said that what-ever her father's quirks—the way he started competitive weight lifting when Yo did, the way he marshaled research when he was feeling emotional, the way he seemed to see his children as charming, lovable home-improvement projects—she always knew that he loved them, that their mother loved them, that Tea, who had been in their lives for many years, loved them. "I just think there was never that much room for error," E said. "Because parenting doesn't matter that much, pro-vided you give your kids love."

E was in love with love, she said, all kinds of love—future love, romantic love, married love, father love, and mother love. She could not wait to have children. Once you have children, she often thought, you would love them absolutely. And since you loved them absolutely, that meant that nothing in your life could ever, ever be counted as a mistake, because everything in your life had led you to them. That test you bombed in eleventh grade? Maybe if you had passed it and been accepted at the University of Michigan, you would never have met the man who was such a jerk that you were ready to appreciate the man you ultimately did marry, who was the father of the children whom you loved exactly as they were. Maybe the bitter fight you and your husband had the day you went to the fertility doctor was perfect, be-cause that meant that you conceived the next round and not that first round, thus giving you the children for whom you could not imagine a more perfect love. Maybe you bumped your hip on a nightstand on your way to the bathroom one evening, which explained why one par-ticular sperm met that egg.

Parenthood would be bliss, total absolution; parenthood, she thought with the innocence of the wholly uninitiated, would mean no mistakes.

Acknowledgments

Sometimes when I can't sleep, I can enter a deep state of calm just by thinking about the people in my life for whom I am grateful, many of whom were essential to the writing of this book. That list in particular must start where the book did, with my agent, Elyse Cheney, a creative force who operates with fierce integrity and who provided crucial editorial guidance, and my editor, Gillian Blake, who read more drafts than any human ever should have to. The steady support of someone whom I admire in so many ways has been a true and exceptional gift, and I can only hope she knows how much I appreciate not just her patience but her superb judgment in all things. Thank you to Jess Scott, Ashley Pierce, Amy Li, and Grace Johnson for their assiduous care in all matters big and small, and to Hilary Roberts and Diana Drew for their eagle-eyed copy edits.

I am immensely grateful to the family members who agreed to participate in this project and to all of the scientists I feature, many of whom tentatively exited their comfort zones of data and statistical analysis to enter, with game generosity, the hazier world of their own family histories. In addition to the researchers quoted in this book, others spent many hours talking to me, explaining concepts and offering big-picture perspectives (and sometimes personal ones). Among them are Brent Roberts, Elliot Tucker-Drob, Larry Steinberg, Susan

McHale, Eric Turkheimer, Jenae Neiderhiser, and Teresa Amabile. The remarkable Adam Grant always lives up to his reputation as an extraordinarily kind and generous point person for perspective, research references, and sage counsel.

Other editors in my life have shared essential thoughts with me for the book, engaged in lively conversation with me about it, or simply given me so much feedback about other work over the years that I have internalized their responses in ways that, I hope, have made me a better writer. Among those are Ilena Silverman, Rachel Poser, Lauren Kern, and Jake Silverstein. Jake functions as a boss in a way that reminds me of many of the parents whom I admire—he is reflexively supportive, which is precisely why his devoted writers are so eager to meet his high standards. The energizing conversations about the topic of this book that I have enjoyed with Carolyn Ryan and Jodi Kantor fortified me at crucial moments and sparked new lines of inquiry, for which I am exceedingly grateful. Aisha Khan's sage feedback reassured me that she could help me try to rectify my blind spots.

One of the best things about being a writer is that you get to know and become friends with other writers and editors. I am beyond lucky that some of the dear friends I've made that way are not only brilliant but were willing to give me essential feedback, often on the turn of a dime, when I asked for it. Thank you, thank you, Ariel Kaminer, Jodi Wilgoren, and Jennifer Senior.

It was a joy to have the eyes of former students, such as Rebecca Salzhauer and Grace Cajski, on some of these chapters. For research assistance, I thank Yinuo Shi and Ella Fanger, as well as Gregory Barber, who was in it for the long haul. Greg's even-keeled good humor, deft touches, and attention to detail provided me great solace over many months. Richard Rampell provided support and wise counsel.

I could not have finished this book any more than I could have survived many years of parenting without the many other friends who have supported me over the years: Fellow retreaters Melanie Hoopes,

Jodi Levine, Molly Roberts, and Melissa Kirsch; the core crew of Eva Kolodner, Jenny Lord, Abigail Smith, and Nan Strauss; my Hastings rocks and rock stars (as mothers and everything else) Jennifer Dunnington and Jana Dolgins; my consigliere and cheerleader, Elizabeth Fishman; and the friends who witnessed my own childhood, as I did theirs, Liz Altman Smoltz and Leslie Platt-Zolov. Being able to call Lisa Damour for insights about parenting is like having Julia Child on speed dial while you're making your first soufflé; Jen Cohen, Paul Kraske, Peter Hyde, and Stacy Swann have possibly added years to my life by teaching me how to laugh, at myself most of all. Felicia Wong, Jon Halperin, Jake Barton, and Jenny Raymond are creative inspirations who beautifully blur the line between friends and family. Anya Epstein and I, former roommates, continued the work of raising each other, I sometimes think; I could not have been in better, steadier, more loving hands.

One of the highlights of this project was having the editorial input of my niece, Lila Dominus; she is one of a host of nieces and nephews (and their loved ones), who suggest the best is yet to come: Coby, Cari, Emily, Olive, Noah, and Erika. My brother Andrew's enormous faith in me has been, at times, daunting, but always loving; in recent years, as my sister and I have embarked on near daily walks, rarely has a thought gone through my head without my sister Ellen knowing of it. I wouldn't share them all if she didn't have so much wisdom to offer in response. Andrew's wife Stephanie and Ellen's husband James round out the kind of supportive and loving family that provides great comfort, as do my husband's siblings: Jan, Dennis, Stephen, and Alex.

I have, of course, thought a good deal about my own parents and upbringing over the course of this book. My parents are people who modeled both industry (our home was spotless and orderly, my father's work hours long) and also the true enjoyment of leisure—of travel, of hanging around in a bathrobe until 10:30 on a Sunday, of enjoying a meal out when possible. In one of my fondest memories of my mother, she is urging me to just turn in a paper I was agonizing over, because who really cared if I got a B? (Only me, apparently.) They remind me of some of the parents in the book, in that they fretted over my decision to be a writer—it didn't even have any clear career track!—even as they had encouraged it when I was young, splurging on the books and the journals I always wanted. Whatever goals I have reached professionally ultimately reflects their good fortune and hard work—both allowed me to graduate from college debt free, which made it easier for me to pursue the work that I have.

For unconditional love, I have them to thank; because I had always had that, I was able to recognize it when it came my way from Alan Burdick, my husband of twenty-one years, a brilliant writer and editor and the love of my life. I have failed as a parent in more ways than I can count; but I know I did right by my children in one key regard, which is in choosing their father. Our sons, Leo and Josh, so different from us, and so different from each other, somehow have in common great kindness and loving patience. They have helped us grow into who we hope to be, by showing us every day what really matters.

Selected Bibliography

It would not take great labors of textual analysis to discern that this book reflects my own idiosyncratic obsessions and is therefore heavy on high-achieving siblings whose ranks include writers, starting with the Brontës. Others who love reading the Brontës—or who simply love reading—will likely share the great pleasure I took in poring over some of the superb books that have been written about them, works to which I am deeply indebted, starting with Juliet Barker's definitive and exhaustive *The Brontës*. The collection of letters that Barker published—*The Brontës: A Life in Letters*—is one of my most prized possessions, and I recommend it to anyone who wants the direct experience of "soul speaking to soul," to rely on a phrase George Henry Lewes used in his review of *Jane Eyre*. Claire Harmon's *A Fiery Heart* captures, unflinchingly, the outrage that fueled Charlotte, a wild talent who felt injustice of all kinds so keenly, and Lucasta Miller's *The Brontë Myth*, an erudite page-turner (my favorite kind), illuminates how the ever-changing reception of the Brontë sisters' work and the interpretations of stories of their lives shift with changing cultural preoccupations.

I have organized the various sources in this book by chapter and endeavored to be clear enough in the body of the work that those sources are readily identifiable in this bibliography. The bulk of this book, however, is the product of countless interviews with family members who tried, honestly and unflinchingly, to answer my many questions about some of the most intimate and personal aspects of their lives, even as those lives were changing over the course of the years during which I was writing this book. I will forever be grateful to them for their patience. They have changed the way I see myself as a parent, but also how I see the world.

INTRODUCTION

Colvin, Geoff. "The World's 50 Greatest Leaders." *Fortune* (March 23, 2017).

Conley, Dalton. *The Pecking Order: A Bold New Look at How Family and Society Determine Who We Become.* New York: Penguin, 2004.

Dominus, Susan. "Table Talk: The New Family Dinner." *New York Times* (April 27, 2012).

Edes, Gordon. "A Dream Born and Fulfilled Early: Theo Epstein, 28, New Red Sox GM." *Boston Globe* (November 26, 2002).

Epstein, Leslie. "How the Epstein Twins Drove Jack Warner Nuts." *Tablet* (February 13, 2017).

Gaskell, Elizabeth. *The Life of Charlotte Brontë.* New York: Penguin Classics, 1998.

Gopnik, Alison. *The Gardener and the Carpenter: What the New Science of Child Development Tells Us About the Relationship Between Parents and Children.* New York: St. Martin's Press, 2017.

Kolhatkar, Sheelah. "The Foer Family." *Observer* (December 18, 2006).

Lincoln, Evelyn. *My Twelve Years with John F. Kennedy.* New York: D. McKay Company, 1965.

Robinson, Mary F. *Emily Brontë.* London: W. H. Allen, 1883.

CHAPTER ONE: THE GROFFS

Berger, Jonah. *Invisible Influence: The Hidden Forces That Shape Behavior.* New York: Simon & Schuster, 2016.

Clarey, Christopher. "The Bryan Brothers Retire as They Played: Together." *New York Times* (August 26, 2020).

"F&M's First Majorette One of Top Twirlers in the US." *Lancaster New Era* (July 14, 1969).

Gopnik, Alison, and Adam Gopnik. "What Our Siblings Do to Us." *New York Times* (September 23, 2011).

Groff, Lauren. *The Monsters of Templeton.* New York: Hachette Books, 2008.

———. "Machado de Assis at the Rio Olympics." *Sewanee Review* 125, no. 1 (Winter 2017): 92–101.

———. *The Vaster Wilds.* New York: Riverhead Books, 2023.

Heinrichs, April, and Matt Robinson. "Finding the Next Mia Hamm and Alex Morgan." *Soccer Journal* (November/December 2014): 64–70.

Hotz, V. Joseph, and Juan Pantano. "Strategic Parenting, Birth Order, and School Performance." *Journal of Population Economics* 28, no. 4 (October 2015): 911–936.

Layden, Tim. "After Rehabilitation, the Best Michael Phelps May Lie Head." *Sports Illustrated* (November 9, 2015).

Matthews, Chris. *Bobby Kennedy: A Raging Spirit.* New York: Simon & Schuster, 2017.

Skipper, Jason. "The Rumpus Interview with Lauren Groff." *Rumpus* (August 24, 2012).

Thomas-Corr, Johanna. "Lauren Groff: 'I Often Get Very Lonely Because My Job Is Very Lonely.'" *Guardian* (September 11, 2021).

Walsh, Mary Roth. *Doctors Wanted: No Women Need Apply.* New Haven and London: Yale University Press, 1977.

Zuckerman, Marvin. "All Parents Are Environmentalists Until They Have Their Second Child." *Behavioral and Brain Sciences* 10, no. 1 (1987): 42–44.

CHAPTER TWO: GENERATORS

Aldeman, Chad. "Much Ado About Grit? An Interview with a Leading Psych Researcher." *Bellwether* (August 4, 2016).

Bell, Katherine. "Life's Work: Wynton Marsalis." *Harvard Business Review* (January/February 2011).

Brontë, Charlotte. *The Letters of Charlotte Brontë.* Edited by Margaret Smith. Oxford, UK: Clarendon Press, 1995.

Brontë, Emily. *Wuthering Heights.* New York: Penguin Classics, 2002.

Duckworth, Angela. "Don't Grade Schools on Grit." *New York Times* (March 27, 2016).

———. *Grit: The Power of Passion and Perseverance.* New York: Scribner, 2016.

Epstein, David. *The Sports Gene: Inside the Science of Extraordinary Athletic Performance.* New York: Portfolio, 2014.

———. *Range: Why Generalists Triumph in a Specialized World.* New York: Riverhead Books, 2019.

Gladwell, Malcolm. *Outliers: The Story of Success.* New York: Little, Brown, and Company, 2008.

Gourse, Leslie. *Wynton Marsalis: Skain's Domain: A Biography.* New York: Shirmer Trade Books, 1999.

Hajdu, David. "Wynton's Blues." *Atlantic* (March 2003).

Harman, Claire. *Charlotte Brontë: A Fiery Heart.* New York: Knopf Doubleday, 2017.

Leonard, Julia A., et al. "Children Persist Less When Adults Take Over." *Child Development* 92, no. 4 (2021): 1325–1336.

————. "Practice What You Preach: How Adults' Actions, Outcomes, and Testimony Affect Preschoolers' Persistence." *Child Development* (September 2019): 1–18.

Leonard, Julia A., Yuna Lee, and Laura E. Schulz. "Infants Make More Attempts to Achieve a Goal When They See Adults Persist." *Science* 357, no. 6357 (September 21, 2017): 1290–1294.

Ritchie, S. J., and E. M. Tucker-Drob. "How Much Does Education Improve Intelligence? A Meta-Analysis." *Psychological Science* 29, no. 8 (2018): 1358–1369.

Saplakoglu, Yasemin. "If at First You Don't Succeed, Show Your Baby Again." *Scientific American* (September 21, 2017).

Shachnai, Reut, et al. "Pointing Out Learning Opportunities Reduces Over-parenting." PsyArXiv (April 3, 2024).

Solomon, Deborah. "The Music Man." *New York Times Magazine* (October 3, 2004).

Tuttle, Christina Clark, et al. "KIPP Middle Schools: Impacts on Achievement and Other Outcomes." *Mathematica Policy Research Reports* (January 1, 2013).

"Wynton Marsalis: Interview." American Academy of Achievement (1991).

CHAPTER THREE: THE HOLIFIELDS

Ashton, Kim. "Siblings in the Struggle." *Harvard Law Bulletin* (Summer 2014).

Bell, Alex, et al. "Who Becomes an Inventor in America? The Importance of Exposure to Innovation." *Quarterly Journal of Economics* 134, no. 2 (May 2019): 647–713.

Bell, Derrick A. "Resignation Letter to the Dean of USC Law Center" (June 2, 1969).

Chartle, Sandra J. "Marilyn Holifield: Breaking Barriers in the Field of Law." *Miami Times* (April 8. 2009).

"City Considers Sale of Pools." *Tallahassee Democrat* (July 6, 1964).

"City Hall Picketed, Open Pools." *Tallahassee Democrat* (June 23, 1966).

Colen, Cynthia G., Nicolo P. Pinchak, and Kierra S. Barnett. "Racial Disparities in Health Among College-Educated African Americans: Can Attendance at Historically Black Colleges or Universities Reduce the Risk of Metabolic Syndrome in Midlife?" *American Journal of Epidemiology* 190, no. 4 (April 2021): 553–561.

Dawson, Kevin. *Undercurrents of Power: Aquatic Culture in the African Diaspora.* Philadelphia: University of Pennsylvania Press, 2018.

Drummond, Tammerlin. "Barack Obama's Law: Harvard Law Review's First Black President Plans a Life of Public Service." *Los Angeles Times* (March 12, 1990).

Due, Tananarive, and Patricia Stephens Due. *Freedom in the Family: A Mother-Daughter Memoir of the Fight for Civil Rights.* New York: Ballantine Books, 2004.

Edwards, Ashley, et al. "HBCU Enrollment and Longer-Term Outcomes." *EdWorkingPaper* 23, no. 883 (December 2023).

Eisenstadt, Marvin, et al. *Parental Loss and Achievement.* Madison, CT: International Universities Press, 1989.

Ensley, Gerald. "Pool Reflected Changing Times." *Tallahassee Democrat* (February 11, 2011).

Felix, Camonghne. "Simone Biles Chooses Herself." *New York* (September 27, 2021).

Fiedler, Tom. "Economic Truth Broke Stalemate." *Miami Herald* (May 13, 1993).

"Five Lincoln Pupils Drown." *Tallahassee Democrat* (May 26, 1946).

Gaskell, Elizabeth. *The Life of Charlotte Brontë.* New York: Penguin Classics, 1998.

Goodman, Joshua, et al. "O Brother, Where Start Thou? Sibling Spillovers in College Enrollment." *NBER Working Papers*, no. 26502 (November 2019).

Gumprecht, Blake. *The American College Town.* Amherst: University of Massachusetts Press, 2008.

Haufler, Arline. "Pool Openers Win in City-Wide Vote." *Tallahassee Democrat* (August 1, 1967).

Holden, Anna, and John Howard Griffin. *Field Reports on Desegregation in the South: Tallahassee, Florida.* New York: Anti-Defamation League of B'nai B'rith, 1956.

Holifield, Edward. "Biomass Plant May Kill More Babies." *Tallahassee Democrat* (November 20, 2008).

Johnson, Dwight, et al. "The Social and Economic Status of the Black Population in the United States: An Historical View 1790–1978." U.S. Department of Commerce, Bureau of the Census. *Current Population Reports* P-23, no. 80 (June 1979).

"Leon's Anchor Club Sets Dance." *The Tallahassee Democrat.* November 30, 1968.

Li, D. J., et al. "Risks of Major Mental Disorders After Parental Death in Children, Adolescents, and Young Adults and the Role of Premorbid Mental Comorbidities: A Population-Based Cohort Study." *Social Psychiatry and Psychiatric Epidemiology* 57, no. 12 (December 2022): 2393–2400.

McCarthy, Joe. *The Remarkable Kennedys.* New York: Popular Library, 1960.

Morris, Eugene. "Millicent Holifield to Be Honored for Work in Nursing School." *Tallahassee Democrat* (May 19, 1990).

Obama, Michelle. *Becoming.* New York: Crown, 2018.

Portman, Jennifer. "TMH Aims for 'Baby-Friendly' Designation." *Tallahassee Democrat* (February 1, 2014).

"Practical Nurses Courses at Lincoln." *Tallahassee Democrat* (October 22, 1957).

Rabby, Glenda Alice. *The Pain and the Promise: The Struggle for Civil Rights in Tallahassee, Florida.* Athens: University of Georgia Press, 1999.

Seymour, Sean, and Julie Ray. "Grads of Historically Black Colleges Have Well-Being Edge." Gallup, 2015.

Shah, Neil H. "What Critical Race Theory Was—and Is—at Harvard Law School." *Harvard Crimson* (May 23, 2023).

Spencer, Merianne R., et al. "Unintentional Drowning Deaths among Children Aged 0–17 Years: United States, 1999–2019." *PubMed* (July 2021).

"The Lesson of Courtney Smith's Death." *Philadelphia Inquirer* (January 17, 1969).

Thyden, N. H., et al. "Family Deaths in the Early Life Course and Their Association with Later Educational Attainment in a Longitudinal Cohort Study." *Social Science and Medicine* 333 (September 2023): 116–161.

Wilkins, David, Elizabeth Chambliss, Lisa A. Jones, and Halie Adamson. "Harvard Law School Report on the State of Black Alumni 1869–2000." Harvard Law School Center on the Legal Profession, 2002.

Williams, Deborah. "Bishop Holifield, Pioneering Florida Conservationist, Dies." *Tallahassee Democrat* (February 2, 1998).

CHAPTER FOUR: EXPECTATIONS

American Psychological Association. "Rising Parental Expectations Linked to Perfectionism in College Students." *ScienceDaily* (March 2022).

Anderson, Jenny. "Parents: Your Absurdly High Expectations Are Harming Your Children's Achievement." *Quartz* (November 26, 2015).

Barker, Juliet. *The Brontës: A Life in Letters.* New York: Abrams Press, 1994.

Briley, D. A., K. Paige Harden, and Elliot M. Tucker-Drob. "Child Characteristics and Parental Educational Expectations: Evidence for Transmission with Transaction." *Developmental Psychology* 50, no. 12 (2014): 2614–2632.

Brontë, Charlotte. *The Letters of Charlotte Brontë.* Edited by Margaret Smith. Oxford, UK: Clarendon Press, 1995.

Gaskell, Elizabeth. *The Life of Charlotte Brontë.* New York: Penguin Classics, 1998.

Green, Dudley. *Patrick Brontë: Father of Genius.* Cheltenham, England: The History Press, 2010.

Loughlin-Presnal, John E., and Karen L. Bierman. "Promoting Parent Academic Expectations Predicts Improved School Outcomes for Low-Income Children Entering Kindergarten." *Journal of School Psychology* 62 (June 2017): 67–80.

Murayama, Kou, et al. "Don't Aim Too High for Your Kids: Parental Overaspiration Undermines Students' Learning in Mathematics." *Journal of Personality and Social Psychology* 111, no. 5 (2016): 766–779.

Rozek, Christopher S., et al. "Utility-Value Intervention with Parents Increases Students' STEM Preparation and Career Pursuit." *Proceedings of the National Academy of Sciences* 114, no. 5 (January 17, 2017): 909–914.

Yeager, David Scott, et al. "Breaking the Cycle of Mistrust: Wise Interventions to Provide Critical Feedback Across the Racial Divide." *Journal of Experimental Psychology: General* 143, no. 2 (2014): 804–824.

CHAPTER FIVE: THE MURGUÍAS

Bruck, Connie. "Ari Emanuel Takes On the World." *New Yorker* (April 19, 2021).

Hagler, Matthew. "Familial Factors Impacting First-Generation College Students' Adjustment: The Compensatory Role of Older Siblings with College Experience." *Journal of First-Generation Student Success* (June 26, 2024): 1–21.

Harris, Kyle. "The Influence of Siblings Who Attended Higher Education on First-Generation College Students." Temple University (2022).

Kuhnhenn, James. "The House and the Hill." *Kansas City Star* (April 30, 1995).

Landon, Jan. "Chain of Support." *Kansas City Star* (February 8, 2004).

Lovell, Mary S. *The Sisters: The Saga of the Mitford Family.* New York: W. W. Norton & Company, 2011.

"Six-year Degree Attainment Rates for Students Enrolled in a Post-Secondary Institute." Pell Institute Fact Sheet (December 14, 2011).

Williams, Carol J. "Obama Nominates Arizona. Judge to Court of Appeals." *Los Angeles Times* (March 27, 2010).

Zaveri, Mihir. "Federal Judge in Kansas City Is Reprimanded for Sexual Harassment." *New York Times* (September 30, 2019).

———. "Federal Judge in Kansas Resigns After Reprimand for Sexual Harassment." *New York Times* (February 19, 2020).

CHAPTER SIX: LUCK AND FATE

Adler, Alfred. *Problems of Neurosis: A Book of Case Histories.* Edited by Philip Mairet. London: Routledge, 1999.

Black, Sandra E., Paul J. Devereux, and Kjell G. Salvanes. "Too Young to Leave the Nest? The Effects of School Starting Age." *Review of Economics and Statistics* 93, no. 2 (May 2011): 455–467.

Black, Sandra E., Erik Grönqvist, and Björn Öckert. "Born to Lead? The Effect of Birth Order on Noncognitive Abilities." *Review of Economics and Statistics* 100, no. 2 (May 2018): 274–286.

Block, Melissa. "U.S. Biathlete Gives Up Olympic Spot to Her Twin Sister." NPR (January 16, 2014).

Broughton, Thomas, et al. "Relative Age in the School Year and Risk of Mental Health Problems in Childhood, Adolescence and Young Adulthood." *Journal of Child Psychology and Psychiatry* 64, no. 1 (August 2022): 185–196.

Bzdek, Vincent. *The Kennedy Legacy.* New York: St. Martin's Press, 2009.

Conley, Dalton. *The Pecking Order: A Bold New Look at How Family and Society Determine Who We Become.* New York: Penguin, 2004.

Damian, R. I., and B. W. Roberts. "Settling the Debate on Birth Order and Personality." *Proceedings of the National Academy of Sciences of the United States of America* 112, no. 46 (2015): 14119–14120.

Dominus, Susan. "The Mixed-Up Brothers of Bogotá." *New York Times* (July 9, 2015).

Frank, Robert H. *Success and Luck: Good Fortune and the Myth of Meritocracy.* Princeton, NJ: Princeton University Press, 2017.

Freese, J., B. Powell, and L. C. Steelman. "Rebel Without a Cause or Effect: Birth Order and Social Attitudes." *American Sociological Review* 64, no. 2 (1999): 207–231.

Freud, Sigmund. "New Introductory Lectures on Psycho-Analysis." *Archives of Neurology and Psychiatry* 33, no. 5 (May 1, 1935): 917–1142.

Freund, Julia, et al. "Emergence of Individuality in Genetically Identical Mice." *Science* 340, no. 6133 (May 9, 2013): 756–759.

Galton, Francis. *English Men of Science: Their Nature and Nurture.* London: Macmillan, 1874.

Gladwell, Malcolm. *Outliers: The Story of Success.* New York: Little, Brown, and Company, 2008.

Gysi, Sabine. "Lifespan Development: The More We Know About It, the More Complex It Seems." *BOLD* (February 17, 2020).

Hotz, V. J., and J. Pantano. "Strategic Parenting, Birth Order, and School Performance." *Journal of Population Economics* 28, no. 4 (October 2015): 911–936.

Karlstad, Øystein, et al. "ADHD Treatment and Diagnosis in Relation to Children's Birth Month: Nationwide Cohort Study from Norway." *Scandinavian Journal of Public Health* 45, no. 4 (May 8, 2017): 343–349.

Kennedy, Rose Fitzgerald. *Times to Remember.* Cutchogue, NY: Buccaneer Books, 1994.

Kohler, Sheila. *Becoming Jane Eyre.* London: Corsair, 2011.

Lejarraga, Tomás, et al. "No Effect of Birth Order on Adult Risk Taking." *Proceedings of the National Academy of Sciences* 116, no. 13 (March 11, 2019): 6019–6024.

Lovell, Mary S. *The Sisters: The Saga of the Mitford Family.* New York: W. W. Norton & Company, 2011.

Oskarsson, Sven, et al. "Big Brother Sees You, but Does He Rule You? The Relationship Between Birth Order and Political Candidacy." *Journal of Politics* 83, no. 3 (May 27, 2021): 1158–62.

Puhani, Patrick A., and Andrea Maria Weber. "Persistence of the School Entry Age Effect in a System of Flexible Tracking." *IZA Discussion Paper,* no. 2935 (2007).

Rohrer, Julia M., Boris Egloff, and Stefan C. Schmukle. "Examining the Effects of Birth Order on Personality." *Proceedings of the National Academy of Sciences* 112, no. 46 (October 19, 2015): 14224–14229.

Segal, Nancy L. *Born Together—Reared Apart.* Cambridge, MA: Harvard University Press, 2012.

———. "Cooperation, Competition, and Altruism Within Twin Sets: A Reappraisal." *Ethology and Sociobiology* 5, no. 3 (January 1984): 163–77.

———. *Deliberately Divided.* Lanham, MD: Rowman & Littlefield, 2021.

Segal, Nancy L., and T. J. Bouchard, Jr. "Grief Intensity Following the Loss of a Twin and Other Relatives: Test of Kinship Genetic Hypotheses." *Human Biology* 65, no. 1 (February 1993): 87–105.

Sharrow, David J., and James J. Anderson. "A Twin Protection Effect? Explaining Twin Survival Advantages with a Two-Process Mortality Model." *PLOS ONE* 11, no. 5 (May 2016).

Sulloway, F. J. *Born to Rebel: Birth Order, Family Dynamics, and Creative Lives.* New York: Pantheon Books, 1996.

Zang, Emma, Poh Lin Tan, and Philip J. Cook. "Sibling Spillovers: Having an Academically Successful Older Sibling May Be More Important for Children in Disadvantaged Families." *American Journal of Sociology* 128, no. 5 (March 1, 2023): 1529–1571.

CHAPTER SEVEN: THE CHENS

Schneider, Gregory. "Two Cities Share a Name, but One Is in Big Trouble." *Washington Post* (July 7, 2018).

CHAPTER EIGHT: FINDING NURTURE

Anderson, Elise L., et al. "The Role of the Shared Environment in College Attainment: An Adoption Study." *Journal of Personality* 89, no. 3 (October 27, 2020).

Kong, Augustine, et al. "The Nature of Nurture: Effects of Parental Genotypes." *Science* 359, no. 6374 (January 25, 2018): 424–428.

Kosse, Fabian, et al. "The Formation of Prosociality: Causal Evidence on the Role of Social Environment." *Journal of Political Economy* 128, no. 2 (February 2020).

Kraft, Matthew A., and Grace T. Falken. "A Blueprint for Scaling Tutoring and Mentoring Across Public Schools." *AERA Open* 7 (2021).

Kraft, Matthew A., et al. "What Impacts Should We Expect from Tutoring at Scale? Exploring Meta-Analytic Generalizability." EdWorkingPaper (2024): 24–1031.

Neville, Helen, et al. "Effects of Music Training on Brain and Cognitive Development in Under-Privileged 3- to 5-Year-Old Children: Preliminary Results." In *Learning, Arts, and the Brain: The Dana Consortium Report on Arts and Cognition,* edited by C. Asbury and B. Rich, 105–116. New York: Dana, 2008.

Nielsen, François and J. Micah Roos. "Genetics of Educational Attainment and the Persistence of Privilege at the Turn of the 21st Century." *Social Forces* 94, no. 2 (June 17, 2015): 535–561.

Resnjanskij, Sven, et al. "Mentoring Improves the School-to-Work Transition of Disadvantaged Adolescents." *VoxEU* (December 17, 2023).

Wolla, Scott A., Guillaume Vandenbroucke, and Cameron Tucker. "Is College Still Worth the High Price? Weighing Costs and Benefits of Investing in Human Capital." *Page One Economics* (September 2023).

CHAPTER NINE: THE PAULUS FAMILY

Barker, Juliet. *The Brontës: A Life in Letters.* New York: Abrams Press, 1994.

Bryer, Jackson R. "An Uncommon Woman: An Interview with Wendy Wasserstein." *Theatre History Studies* 29, no. 1 (2009): 1–17.

Gilbreth, Frank B., and Ernestine Gilbreth Carey. *Cheaper by the Dozen.* New York: Harper Perennial, 2019.

Gourse, Leslie. *Wynton Marsalis: Skain's Domain: A Biography.* New York: Shirmer Trade Books, 1999.

Hardwick, Elizabeth. "Working Girls: The Brontës." *New York Review of Books* (May 4, 1972).

Isherwood, Charles. "Swimming with Watermelons." *Variety* (April 11, 2002).

Klein, Julia. "Out of the Woods." *Columbia Magazine* (Spring 2012):36–41.

Krementz, Jill. *A Very Young Dancer.* New York: Dell Publishing Company, 1986.

Kushner, Tony. "Notes about Political Theater." *Kenyon Review* 19, Issues 3–4 (Summer/Fall 1997): 19–34.

Lahr, John. "After Angels." *New Yorker* (December 20, 2004): 183–4.

Lublin, Joann S. *Earning It: Hard-Won Lessons from Trailblazing Women at the Top of the Business World.* New York: Harper Business, 2016.

Marks, Peter. "They Be Foolish Mortals Who Love the Nightlife." *New York Times* (August 27, 1999).

Needham, Alex. "Tony Kushner on Spielberg, Ye, and 'The Orange-Covered Mud Devil.'" *Guardian* (December 16, 2022).

"New Met Opera Opens." *New York Herald Tribune* (September 19, 1966).

Paulus, Diane, and Randy Weiner. *The Donkey Show: A Midsummer Night's Disco.* Club El Flamingo, 1999.

Shephard, Richard F. "Children Do Musical by Swados." *New York Times* (February 16, 1979).

Taylor, Sydney. *All-of-a-Kind Family.* New York: Yearling, 1984.

Wasserstein, Wendy. "The Baby Who Arrived Too Soon." *New Yorker* (February 13, 2000): 87–109.

Wilderotter, Maggie, and Denise Morrison. "The Art of Responsibility." TED, 2018.

CHAPTER TEN: OPENNESS

Alan, Sule, and Ipek Mumcu. "Nurturing Childhood Curiosity to Enhance Learning: Evidence from a Randomized Pedagogical Intervention." *American Economic Review* 114, no. 4 (April 2024): 1173–1210.

Allik, Jüri, Anu Realo, and Robert R. McCrae. "Conceptual and Methodological Issues in the Study of the Personality-and-Culture Relationship." *Frontiers in Psychology* 14 (2023): 107–7851.

Baer, Susan. "Brothers: Rahm Emanuel and His Family." *Washingtonian* (May 1, 2008).

Emanuel, Ezekiel. *Brothers Emanuel: A Memoir of an American Family.* New York: Random House, 2013.

Habegger, Alfred. *The Father: A Life of Henry James, Sr.* New York: Farrar, Straus & Giroux, 1994.

James, William. *The Principles of Psychology.* New York: Cosimo Classics, 1890.

Lechner, Clemens M., Ai Miyamoto, and Thomas Knopf. "Should Students Be Smart, Curious, or Both? Fluid Intelligence, Openness, and Interest Co-Shape the Acquisition of Reading and Math Competence." *Intelligence* 76 (2019): 101–378.

Malanchini, Margherita, et al. "'Same but Different': Associations Between Multiple Aspects of Self-Regulation, Cognition, and Academic Abilities." *Journal of Personality and Social Psychology* 117, no. 6 (December 2019): 1164–1188.

McCrae, Robert. "Aesthetic Chills as a Universal Marker of Openness to Experience." *Motivation and Emotion* 31, no. 1 (February 2007): 5–11.

McCrae, Robert, and Paul Costa. "Conceptions and Correlates of Openness to Experience." In *Handbook of Personality Psychology*, edited by Robert Hogan, John Johnson, and Stephen Briggs, 825–847. San Diego: Academic Press, 1998.

Mischel, Walter. *Personality and Assessment*. New York: Wiley, 1968.

———. "Toward a Cognitive Social Learning Reconceptualization of Personality." *Psychological Review* 80, no. 4 (1973): 252–283.

Niehoff, Esther, Linn Petersdotter, and Philipp Alexander Freund. "International Sojourn Experience and Personality Development: Selection and Socialization Effects of Studying Abroad and the Big Five." *Personality and Individual Differences* 112 (July 2017): 55–61.

Pickett, Jennifer, et al. "Concurrent and Lagged Effects of Counterdispositional Extraversion on Vitality." *Journal of Research in Personality* 87 (August 2020): 103–965.

Richter, Julia, et al. "Do Sojourn Effects on Personality Trait Changes Last? A Five-Year Longitudinal Study." *European Journal of Personality* 35, no. 3 (June 1, 2024).

Strous, Jean. *Alice James: A Biography.* New York: New York Review Books Classics, 2011.

Thrailkill, Jane. *Philosophical Siblings: Varieties of Playful Experience in Alice, William, and Henry James.* Philadelphia: University of Pennsylvania Press, 2022.

van Allen, Zachary M., and John M. Zelenski. "Testing Trait-State Isomorphism in a New Domain: An Exploratory Manipulation of Openness to Experience." *Frontiers in Psychology* 9 (October 15, 2018).

William, James. *The Letters of William James, Vol. 1.* Boston: The Atlantic Monthly Press, 1920.

Zimmermann, Julia, and Franz J. Neyer. "Do We Become a Different Person When Hitting the Road? Personality Development of Sojourners." *Journal of Personality and Social Psychology* 105, no. 3 (2013): 515–530.

CHAPTER ELEVEN: THE WOJCICKIS

Alter, Charlotte. "A Mother's Expert Tips on Raising Confident Daughters." *Time* (August 29, 2016).

Altman, Sam. "Anne Wojcicki on How to Build the Future." *Y Combinator* (April 13, 2018).

Amidzic, O., et al. "Pattern of Focal Gamma-Bursts in Chess Players." *Nature* 412, no. 6847 (August 9, 2001): 603.

Bandura, Albert. "Influence of Models' Reinforcement Contingencies on the Acquisition of Imitative Responses." *Journal of Personality and Social Psychology* 1, no. 6 (June 1965): 589–595.

Brenøe, Anne Ardila. "Brothers Increase Women's Gender Conformity." *Journal of Population Economics* 35 (2021).

Brontë, Charlotte. *The Letters of Charlotte Brontë*. Edited by Margaret Smith. Oxford, UK: Clarendon Press, 1995.

Bryan, Wayne. *Raising Your Child to Be a Champion in Athletics, Arts and Academics.* New York: Citadel Press, 2004.

Colfax, David, and Micki Colfax. *Hard Times in Paradise.* New York: Grand Central Publishing. 1992.

———. *Homeschooling for Excellence.* New York: Grand Central Publishing, 1988.

Dominus, Susan. "Dangerous When Interested." *New York Times* (August 19, 2007).

Flora, Carlin. "The Grandmaster Experiment." *Psychology Today* (July 2005).

Galanaki, Evangelia, and Konstantinos D. Malafantis. "Albert Bandura's Experiments on Aggression Modeling in Children: A Psychoanalytic Critique." *Frontiers in Psychology* 13 (2022): 988877.

Galton, Francis. *Hereditary Genius.* London: Macmillan, 1869.

Gupta, Priya, et al. "A Genome-Wide Investigation Into the Underlying Genetic Architecture of Personality Traits and Overlap with Psychopathology." *Nature Human Behaviour* 8 (Nov. 2024): 2235–2249.

Laporte, Nicole. "Three Things You Need to Know About Susan Wojcicki." *Fast Company* (August 8, 2014).

Lee, James J., et al. "Gene Discovery and Polygenic Prediction from a Genome-Wide Association Study of Educational Attainment in 1.1 Million Individuals." *Nature Genetics* 50, no. 8 (July 23, 2018): 1112–1121.

Okbay, Aysu, et al. "Polygenic Prediction of Educational Attainment Within and Between Families from Genome-Wide Association Analyses in 3 Million Individuals." *Nature Genetics* 54, no. 4 (March 31, 2022): 437–449.

Polgár, László. *Raise a Genius! (Nevelj zsenit!)*. Budapest: Self-published, 1989.

Polgar, Susan, and Paul Truong. *Breaking Through: How the Polgar Sisters Changed the Game of Chess.* London: Gloucester Publishing, 2005.

Rubio, Silvia. "A Back-to-the-Land Family That Never Lost the Faith." *San Francisco Chronicle* (July 21, 1992).

Ryan, Kevin J. "23andMe's Anne Wojcicki Says Doing These Two Things as a Leader Built Her Company's Culture of Honesty." *Inc. Magazine* (March/April 2019):40–45.

Shetty, Jay. "Susan Wojcicki on What It's Like Being the CEO of YouTube and How to Avoid Burnout." Jay Shetty Podcast (October 2022).

Turley, Patrick, et al. "Problems with Using Polygenic Scores to Select Embryos." *Obstetrical & Gynecological Survey* 76, no. 10 (October 2021): 609–610.

Wojcicki, Esther. *How to Raise Successful People: Simple Lessons for Radical Results.* Boston: Mariner Books, 2019.

Wojcicki, Stanley. "Scientists Set to Zero In on Basic Truths About Matter." *Bismarck Tribune* (September 13, 1985).

CHAPTER TWELVE: BETTER, BETTER, BEST

Albert, Hana. "These Siblings Have NYC's Longest and Shortest Names." *New York Post* (May 6, 2018).

Conley, Dalton.

———. "Parent Like a Mad Scientist." *Time* (March 26, 2014).

———. *Parentology: Everything You Wanted to Know about the Science of Raising Children but Were Too Exhausted to Ask.* New York: Simon & Schuster, 2014.

The Pecking Order: A Bold New Look at How Family and Society Determine Who We Become. New York: Penguin, 2004.

———. *The Social Genome: The New Science of Nature and Nurture.* New York: W. W. Norton & Company, 2025.

———. "Stalkers, Twins, and the Case of the Missing Heritability." TEDxUNC, 2015.

Conley, Dalton, et al. "Heritability and the Equal Environments Assumption: Evidence from Multiple Samples of Misclassified Twins." *Behavioral Genetics* 43, vol. 5 (September 2013): 415–426.

Duis, J., and M. G. Butler. "Syndromic and Nonsyndromic Obesity: Underlying Genetic Causes in Humans." *Advanced Biology* 6 (2022): 2101154.

Kim, Ji-Yeon, et al. "Longitudinal Course and Family Correlates of Sibling Relationships from Childhood Through Adolescence." *Child Development* 77, no. 6 (November/December 2006): 1746–1761.

Marchese, David. "Mo Willems Has a Message for Parents: He's Not on Your Side." *New York Times* (November 13, 2020).

Polderman, Tinca J. C., et al. "Meta-Analysis of the Heritability of Human Traits Based on Fifty Years of Twin Studies." *Nature Genetics* 47, no. 7 (May 18, 2015): 702–709.

About the Author

SUSAN DOMINUS has worked for *The New York Times* since 2007, first as a Metro columnist and then as staff writer for *The New York Times Magazine.* In 2018, she was part of a team that reported on workplace sexual harassment issues and won a Pulitzer Prize for public service. She won a Front Page Award from the Newswomen's Club of New York and a Mychal Judge Heart of New York Award from the New York Press Club. She is a lecturer at Yale College and has studied as a fellow at the National Institutes of Health and Yale Law School. Her article about menopause in *The New York Times Magazine* won a National Magazine Award in 2024.